PLASTERING

LEVEL 1

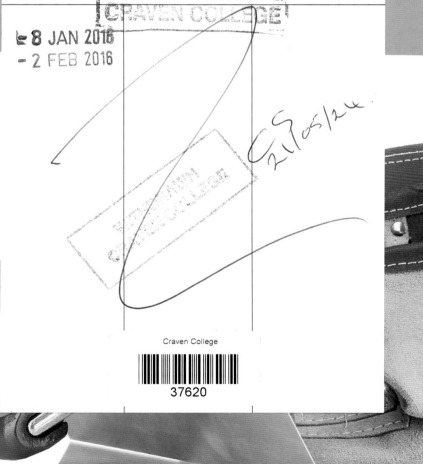

B|A
C|H
BRITISH ASSOCIATION
OF
CONSTRUCTION HEADS

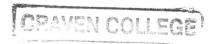

OXFORD
UNIVERSITY PRESS

OXFORD
UNIVERSITY PRESS

Great Clarendon Street, Oxford, OX2 6DP, United Kingdom

Oxford University Press is a department of the University of Oxford. It furthers the University's objective of excellence in research, scholarship, and education by publishing worldwide. Oxford is a registered trade mark of Oxford University Press in the UK and in certain other countries

British Library Cataloguing in Publication Data
Data available

978-1-40-852698-9

10 9 8 7 6 5 4 3 2 1

MIX
Paper from responsible sources
FSC® C007785

Paper used in the production of this book is a natural, recyclable product made from wood grown in sustainable forests. The manufacturing process conforms to the environmental regulations of the country of origin.

Printed in Great Britain by Bell and Bain Ltd., Glasgow.

Acknowledgements

The publishers would like to thank the following for permissions to use their photographs:

ajt/iStock: 6.6; © Adrian Sherratt/Alamy: 5.40; © Anton Starikov/Alamy: 7.2; © Arcaid Images/Alamy: 3.12; © blickwinkel/Alamy: 3.13; © Justin Kase ztwoz/Alamy: 4.26; © Peter Davey/Alamy: 1.0; © Steve Atkins Photography/Alamy: 4.5; © The National Trust Photolibrary/Alamy: 6.1; © vast natalia/Alamy: 7.17; © worldthroughthelens-DIY/Alamy: 7.11; Catnic: 5.60; Chotewang/Shutterstock: 3.20; DGLowrie/iStock: 3.19; Energy Saving Trust © 2011 - marketing@est.org.uk: 3.28; ferrantraite/iStock: 3.26; Filip Ristevski/Shutterstock: 4.14; Fotolia: 1.1, 1.2, 1.3, 1.5, 1.6, 1.7, 1.8, 1.14, 1.15, 1.16; Gary Ombler/Thinkstock: 4.30; Hipped end roof. Chicago_bungalow[1] (Wikipedia): 3.15; ictor/iStock: 3.22; JasonDoiy/iStock: 2.26; Katarzyna Wojtasik/Shutterstock: 5.3; katylh /iStock: 3.27; KjellBrynildsen/

iStock: 3.16; Monkey Business Images/Shutterstock: 3.0; ndoeljindoel/Shutterstock: 2.0; Nelson Thornes: 1.9, 1.10, 1.12, 1.13; ofbeautifulthings/iStock: 3.18; Pavel L Photo and Video/Shutterstock: 3.23; pejft/iStock: 3.25; PETER GARDINER/SCIENCE PHOTO LIBRARY: 1.4; photolia/iStock: 1.11; quaximo/Fotolia: 4.4; Racatac: 7.1; Reproduced by kind permission of Rebecca Nestingen: 4.28; Richard Wilson/Oxford University Press: 4.0, 4.1, 4.2, 4.6, 4.7, 4.8, 4.9, 4.10, 4.11, 4.12, 4.13, 4.15, 4.17, 4.21, 4.22, 4.25, 4.27, 4.31, 4.32, 4.33, 4.35, 4.36, 4.37, 4.38, 4.39, 4.40, 4.41, 4.42, 4.43, 4.44, 4.45, 4.46, 4.47, 4.48, 5.0, 5.4, 5.5, 5.6, 5.7, 5.8, 5.9, 5.10, 5.11, 5.12, 5.13, 5.14, 5.15, 5.16, 5.17, 5.18, 2.19, 5.20, 5.21, 5.22, 5.23, 5.24, 5.25, 5.26, 5.27, 5.28, 5.29, 5.30, 5.31, 5.32, 5.33, 5.34, 5.35, 5.37, 5.38, 5.39, 5.41, 5.42, 5.43, 5.44, 5.50, 5.51, 5.52, 5.53, 5.56, 5.57, 5.58, 5.59, 5.61, 5.62, 5.63, 5.64, 5.65, 5.66, 5.67, 5.68, 5.71, 5.69, 5.70, 5.73, 5.74, 5.72, 5.75, 5.76, 5.77, 5.78, 5.79, 5.80, 5.81, 5.82, 5.85, 5.83, 5.84, 5.86, 5.87, 5.88, 5.89, 5.90, 5.91, 5.92, 5.93, 5.94, 5.95, 5.96, 5.97, 5.98, 5.99, 5.100, 6.0, 6.2, 6.3, 6.4, 6.7, 6.8, 6.9, 6.10, 6.13, 6.16, 6.17, 6.18, 6.19, 6.21, 6.20, 6.22, 6.24, 6.25, 6.26, 6.27, 6.29, 6.30, 6.31, 6.32, 6.33, 6.34, 6.35, 6.36, 6.37, 6.38, 6.39, 6.40, 6.41, 6.42, 6.43, 6.44, 6.47, 6.45, 6.46, 6.48, 6.49, 6.50, 6.51, 6.52, 6.53, 6.54, 6.55, 6.56, 6.57, 6.58, 6.59, 6.60, 6.61, 6.62, 6.63, 6.64, 7.0, 7.3, 7.16, 7.18, 7.23, 7.24, 7.25, 7.26, 7.27, 7.28, 7.29, 7.30; richsouthwales/Shutterstock: 3.14; Ridofranz/iStock: 4.3; small_frog/iStock: 3.4; Stocksnapper/Shutterstock: 6.14; Susan Law Cain/Shutterstock: 3.11; titine974/iStock: 3.21; TTS Tooltechnic Systems GB Ltd: 4.16; Willow Creek Paving Stones: 7.4; www.contenti.com: 6.15; www.fine-tools.com: 6.5; www.tanks-direct.co.uk: 6.23; YK/Shutterstock: 3.24; Yuliya Evstratenko/Shutterstock: 5.36

Although we have made every effort to trace and contact all copyright holders before publication this has not been possible in all cases. If notified, the publisher will rectify any errors or omissions at the earliest opportunity.

Links to third party websites are provided by Oxford in good faith and for information only. Oxford disclaims any responsibility for the materials contained in any third party website referenced in this work.

Note to learners and tutors

This book clearly states that a risk assessment should be undertaken and the correct PPE worn for the particular activities before any practical activity is carried out. Risk assessments were carried out before photographs for this book were taken and the models are wearing the PPE deemed appropriate for the activity and situation. This was correct at the time of going to print. Colleges may prefer that their learners wear additional items of PPE not featured in the photographs in this book and should instruct learners to do so in the standard risk assessments they hold for activities undertaken by their learners. Learners should follow the standard risk assessments provided by their college for each activity they undertake which will determine the PPE they wear.

CONTENTS

INTRODUCTION

About this book

This book has been written for the Cskills Awards Level 1 Diploma in Plastering. It covers all the units of the qualification, so you can feel confident that your book fully covers the requirements of your course.

This book contains a number of features to help you acquire the knowledge you need. It also demonstrates the practical skills you will need to master to successfully complete your qualification. We've included additional features to show how the skills and knowledge can be applied to the workplace, as well as tips and advice on how you can improve your chances of gaining employment.

The features include:

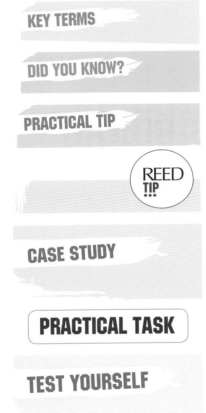

* chapter openers which list the learning outcomes you must achieve in each unit

* key terms that provide explanations of important terminology that you will need to know and understand

* Did you know? margin notes to provide key facts that are helpful to your learning

* practical tips to explain facts or skills to remember when undertaking practical tasks

* Reed tips to offer advice about work, building your CV and how to apply the skills and knowledge you have learnt in the workplace

* case studies that are based on real tradespeople who have undertaken apprenticeships and explain why the skills and knowledge you learn with your training provider are useful in the workplace

* practical tasks that provide step-by-step directions and illustrations for a range of projects you may do during your course

* Test yourself multiple choice questions that appear at the end of each unit to give you the chance to revise what you have learnt and to practise your assessment (your tutor will give you the answers to these questions).

Further support for this book can be found at our website, www.planetvocational.com/subjects/build

CONTRIBUTORS TO THIS BOOK

British Association of Construction Heads

The British Association of Construction Heads is an association formed largely from those managing and delivering the construction curriculum from pre-apprenticeship to post graduate level. The Association is a voluntary organisation and was formed in 1983 and has grown to a position where it can demonstrate that BACH members now manage over 90% of the Learners studying the construction curriculum and includes membership of 80% of the Colleges offering the Construction curriculum in England, Northern Ireland, Scotland and Wales. It accepts membership applications from Colleges and other organisations who are passionate about quality and standards in construction education and training. Visit www.bach.uk.com for more information.

A huge thank you to Mike Morson at Riverside College and David Kehoe at Vision West Nottinghamshire College for their technical expertise in reviewing, advising and facilitating the photo shoot.

Reed Property & Construction

Reed Property & Construction specialises in placing staff at all levels, in both temporary and permanent positions, across the complete lifecycle of the construction process. Our consultants work with most major construction companies in the UK and our clients are involved with the design, build and maintenance of infrastructure projects throughout the UK.

Expert help
As a leading recruitment consultancy for mid–senior level construction staff in the UK, Reed Property & Construction is ideally placed to advise new workers entering the sector, from building a CV to providing expertise and sharing our extensive sector knowledge with you. That's why you will find helpful hints from our highly experienced consultants, designed to help you find that first step on the construction career ladder. These tips range from advice on CV writing to interview tips and techniques, and are linked with the learning material in this book.

Work-related advice
Reed Property & Construction has gained insights from some of our biggest clients to help you understand the mind-set of potential employers. This includes the traits and skills that they would like to see in their new employees, why you need the skills taught in this book and how they are used on a day to day basis within their organisations.

Getting your first job
This invaluable information is not available anywhere else and is geared to helping you gain a position with an employer once you've completed your studies. Entry level positions are not usually offered by recruitment companies, but our advice will help you to apply for jobs in construction and hopefully gain your first position as a skilled worker.

CONTRIBUTORS TO THIS BOOK

The case studies in this book feature staff from Laing O'Rourke and South Tyneside Homes.

Laing O'Rourke is an international engineering company that constructs large-scale building projects all over the world. Originally formed from two companies, John Laing (founded in 1848) and R O'Rourke and Son (founded in 1978) joined forces in 2001.

At Laing O'Rourke, there is a strong and unique apprenticeship programme. It runs a four-year 'Apprenticeship Plus' scheme in the UK, combining formal college education with on-the-job training. Apprentices receive support and advice from mentors and experienced tradespeople, and are given the option of three different career pathways upon completion: remaining on site, continuing into a further education programme, or progressing into supervision and management.

The company prides itself on its people development, supporting educational initiatives and investing in its employees. Laing O'Rourke believes in collaboration and teamwork as a path to achieving greater success, and strives to maintain exceptionally high standards in workplace health and safety.

South Tyneside Council's
Housing Company

South Tyneside Homes was launched in 2006, and was previously part of South Tyneside Council. It now works in partnership with the council to repair and maintain 18,000 properties within the borough, including delivering parts of the Decent Homes Programme.

South Tyneside Homes believes in putting back into the community, with 90 per cent of its employees living in the borough itself. Equality and diversity, as well as health and wellbeing of staff, is a top priority, and it has achieved the Gold Status Investors in People Award.

South Tyneside Homes is committed to the development of its employees, providing opportunities for further education and training and great career paths within the company – 80 per cent of its management team started as apprentices with the company. As well as looking after its staff and their community, the company looks after the environment too, running a renewable energy scheme for council tenants in order to reduce carbon emissions and save tenants money.

The apprenticeship programme at South Tyneside Homes has been recognised nationally, having trained over 80 young people in five main trade areas over the past six years. One of the UK's Top 100 Apprenticeship Employers, it is an Ambassador on the panel of the National Apprentice Service. It has won the Large Employer of the Year Award at the National Apprenticeship Awards and several of its apprentices have been nominated for awards, including winning the Female Apprentice of the Year for the local authority.

Chapter 1

Unit CSA–L1Core01
HEALTH, SAFETY AND WELFARE IN CONSTRUCTION AND ASSOCIATED INDUSTRIES

LEARNING OUTCOMES

LO1: Know the health and safety regulations, roles and responsibilities

LO2: Know the accident and emergency procedures and how to report them

LO3: Know how to identify hazards on construction sites

LO4: Know about health and hygiene in a construction environment

LO5: Know how to handle and store materials and equipment safely

LO6: Know about basic working platforms and access equipment

LO7: Know how to work safely around electricity in a construction environment

LO8: Know how to use personal protective equipment (PPE) correctly

LO9: Know the fire and emergency procedures

LO10: Know about signs and safety notices

INTRODUCTION

The aim of this chapter is to:

* help you to source relevant safety information
* help you to use the relevant safety procedures at work.

HEALTH AND SAFETY REGULATIONS, ROLES AND RESPONSIBILITIES

The construction industry can be dangerous, so keeping safe and healthy at work is very important. If you are not careful, you could injure yourself in an accident or perhaps use equipment or materials that could damage your health. Keeping safe and healthy will help ensure that you have a long and injury-free career.

Although the construction industry is much safer today than in the past, more than 2,000 people are injured and around 50 are killed on site every year. Many others suffer from long-term ill-health such as deafness, spinal damage, skin conditions or breathing problems.

Key health and safety legislation

Laws have been created in the UK to try to ensure safety at work. Ignoring the rules can mean injury or damage to health. It can also mean losing your job or being taken to court.

The two main laws are the Health and Safety at Work etc. Act (HASAWA) and the Control of Substances Hazardous to Health Regulations (COSHH).

The Health and Safety at Work etc. Act (HASAWA) (1974)
This law applies to all working environments and to all types of worker, sub-contractor, employer and all visitors to the workplace. It places a duty on everyone to follow rules in order to ensure health, safety and welfare. Businesses must manage health and safety risks, for example by providing appropriate training and facilities. The Act also covers first aid, accidents and ill health.

Reporting of Injuries, Diseases and Dangerous Occurrences Regulations (RIDDOR) (1995)
Under RIDDOR, employers are required to report any injuries, diseases or dangerous occurrences to the Health and Safety Executive (HSE). The regulations also state the need to maintain an accident book.

KEY TERMS

HASAWA

– the Health and Safety at Work etc. Act outlines your and your employer's health and safety responsibilities.

COSHH

– the Control of Substances Hazardous to Health Regulations are concerned with controlling exposure to hazardous materials.

DID YOU KNOW?

In 2011 to 2012, there were 49 fatal accidents in the construction industry in the UK. (*Source* HSE, www.hse.gov.uk)

KEY TERMS

HSE

– the Health and Safety Executive, which ensures that health and safety laws are followed.

Accident book

– this is required by law under the Social Security (Claims and Payments) Regulations 1979. Even minor accidents need to be recorded by the employer. For the purposes of RIDDOR, hard copy accident books or online records of incidents are equally acceptable.

Control of Substances Hazardous to Health (COSHH) (2002)

In construction, it is common to be exposed to substances that could cause ill health. For example, you may use oil-based paints or preservatives, or work in conditions where there is dust or bacteria.

Employers need to protect their employees from the risks associated with using hazardous substances. This means assessing the risks and deciding on the necessary precautions to take.

Any control measures (things that are being done to reduce the risk of people being hurt or becoming ill) have to be introduced into the workplace and maintained; this includes monitoring an employee's exposure to harmful substances. The employer will need to carry out health checks and ensure that employees are made aware of the dangers and are supervised.

Control of Asbestos at Work Regulations (2012)

Asbestos was a popular building material in the past because it was a good insulator, had good fire protection properties and also protected metals against corrosion. Any building that was constructed before 2000 is likely to have some asbestos. It can be found in pipe insulation, boilers and ceiling tiles. There is also asbestos cement roof sheeting and there is a small amount of asbestos in decorative coatings such as Artex.

Asbestos has been linked with lung cancer, other damage to the lungs and breathing problems. The regulations require you and your employer to take care when dealing with asbestos:

* You should always assume that materials contain asbestos unless it is obvious that they do not.

* A record of the location and condition of asbestos should be kept.

* A risk assessment should be carried out if there is a chance that anyone will be exposed to asbestos.

The general advice is as follows:

* Do not remove the asbestos. It is not a hazard unless it is removed or damaged.

* Remember that not all asbestos presents the same risk. Asbestos cement is less dangerous than pipe insulation.

* Call in a specialist if you are uncertain.

Provision and Use of Work Equipment Regulations (PUWER) (1998)

PUWER concerns health and safety risks related to equipment used at work. It states that any risks arising from the use of equipment must either be prevented or controlled, and all suitable safety measures must have been taken. In addition, tools need to be:

* suitable for their intended use

* safe

REED TIP
...

Employers will want to know that you understand the importance of health and safety. Make sure you know the reasons for each safe working practice.

* well maintained

* used only by those who have been trained to do so.

Manual Handling Operations Regulations (1992)

These regulations try to control the risk of injury when lifting or handling bulky or heavy equipment and materials. The regulations state as follows:

* Hazardous manual handling should be avoided if possible.

* An assessment of hazardous manual handling should be made to try to find alternatives.

* You should use mechanical assistance where possible.

* The main idea is to look at how manual handling is carried out and finding safer ways of doing it.

Personal Protection at Work Regulations (PPE) (1992)

This law states that employers must provide employees with personal protective equipment (PPE) at work whenever there is a risk to health and safety. PPE needs to be:

* suitable for the work being done

* well maintained and replaced if damaged

* properly stored

* correctly used (which means employees need to be trained in how to use the PPE properly).

Work at Height Regulations (2005)

Whenever a person works at any height there is a risk that they could fall and injure themselves. The regulations place a duty on employers or anyone who controls the work of others. This means that they need to:

* plan and organise the work

* make sure those working at height are **competent**

* assess the risks and provide appropriate equipment

* manage work near or on fragile surfaces

* ensure equipment is inspected and maintained.

In all cases the regulations suggest that, if it is possible, work at height should be avoided. Perhaps the job could be done from ground level? If it is not possible, then equipment and other measures are needed to prevent the risk of falling. When working at height measures also need to be put in place to minimise the distance someone might fall.

KEY TERMS

PPE

– personal protective equipment can include gloves, goggles and hard hats.

Competent

– to be competent an organisation or individual must have:

* sufficient knowledge of the tasks to be undertaken and the risks involved

* the experience and ability to carry out their duties in relation to the project, to recognise their limitations and take appropriate action to prevent harm to those carrying out construction work, or those affected by the work.

(*Source* HSE)

Figure 1.1 Examples of personal protective equipment

Employer responsibilities under HASAWA

HASAWA states that employers with five or more staff need their own health and safety policy. Employers must assess any risks that may be involved in their workplace and then introduce controls to reduce these risks. These risk assessments need to be reviewed regularly.

Employers also need to supply personal protective equipment (PPE) to all employees when it is needed and to ensure that it is worn when required.

Specific employer responsibilities are outlined in Table 1.1.

Employee responsibilities under HASAWA

HASAWA states that all those operating in the workplace must aim to work in a safe way. For example, they must wear any PPE provided and look after their equipment. Employees should not be charged for PPE or any actions that the employer needs to take to ensure safety.

Specific employer responsibilities are outlined in Table 1.1. Table 1.2 identifies the key employee responsibilities.

KEY TERMS

Risk

– the likelihood that a person may be harmed if they are exposed to a hazard.

Hazard

– a potential source of harm, injury or ill-health.

Near miss

– any incident, accident or emergency that did not result in an injury but could have done so.

Employer responsibility	Explanation
Safe working environment	Where possible all potential risks and hazards should be eliminated.
Adequate staff training	When new employees begin a job their induction should cover health and safety. There should be ongoing training for existing employees on risks and control measures.
Health and safety information	Relevant information related to health and safety should be available for employees to read and have their own copies.
Risk assessment	Each task or job should be investigated and potential risks identified so that measures can be put in place. A risk assessment and method statement should be produced. The method statement will tell you how to carry out the task, what PPE to wear, equipment to use and the sequence of its use.
Supervision	A competent and experienced individual should always be available to help ensure that health and safety problems are avoided.

Table 1.1 Employer responsibilities under HASAWA

Employee responsibility	Explanation
Working safely	Employees should take care of themselves, only do work that they are competent to carry out and remove obvious hazards if they are seen.
Working in partnership with the employer	Co-operation is important and you should never interfere with or misuse any health and safety signs or equipment. You should always follow the site rules.
Reporting hazards, near misses and accidents correctly	Any health and safety problems should be reported and discussed, particularly a near miss or an actual accident.

Table 1.2 Employee responsibilities under HASAWA

Health and Safety Executive

The Health and Safety Executive (HSE) is responsible for health, safety and welfare. It carries out spot checks on different workplaces to make sure that the law is being followed.

HSE inspectors have access to all areas of a construction site and can also bring in the police. If they find a problem then they can issue an **improvement notice**. This gives the employer a limited amount of time to put things right.

In serious cases, the HSE can issue a **prohibition notice**. This means all work has to stop until the problem is dealt with. An employer, the employees or **sub-contractors** could be taken to court.

The roles and responsibilities of the HSE are outlined in Table 1.3.

Responsibility	Explanation
Enforcement	It is the HSE's responsibility to reduce work-related death, injury and ill health. It will use the law against those who put others at risk.
Legislation and advice	The HSE will use health and safety legislation to serve improvement or prohibition notices or even to prosecute those who break health and safety rules. Inspectors will provide advice either face-to-face or in writing on health and safety matters.
Inspection	The HSE will look at site conditions, standards and practices and inspect documents to make sure that businesses and individuals are complying with health and safety law.

Table 1.3 HSE roles and responsibilities

Sources of health and safety information

There is a wide variety of health and safety information. Most of it is available free of charge, while other organisations may make a charge to provide information and advice. Table 1.4 outlines the key sources of health and safety information.

Source	Types of information	Website
Health and Safety Executive (HSE)	The HSE is the primary source of work-related health and safety information. It covers all possible topics and industries.	www.hse.gov.uk
Construction Industry Training Board (CITB)	The national training organisation provides key information on legislation and site safety.	www.citb.co.uk
British Standards Institute (BSI)	Provides guidelines for risk management, PPE, fire hazards and many other health and safety-related areas.	www.bsigroup.com
Royal Society for the Prevention of Accidents (RoSPA)	Provides training, consultancy and advice on a wide range of health and safety issues that are aimed to reduce work related accidents and ill health.	www.rospa.com
Royal Society for Public Health (RSPH)	Has a range of qualifications and training programmes focusing on health and safety.	www.rsph.org.uk

Table 1.4 Health and safety information

Informing the HSE

The HSE requires the reporting of:

* deaths and injuries – any **major injury**, **over 7-day injury** or death

* occupational disease

* dangerous occurrence – a collapse, explosion, fire or collision

* gas accidents – any accidental leaks or other incident related to gas.

Enforcing guidance

Work-related injuries and illnesses affect huge numbers of people. According to the HSE, 1.1 million working people in the UK suffered from a work-related illness in 2011 to 2012. Across all industries, 173 workers were killed, 111,000 other injuries were reported and 27 million working days were lost.

The construction industry is a high risk one and, although only around 5 per cent of the working population is in construction, it accounts for 10 per cent of all major injuries and 22 per cent of fatal injuries.

The good news is that enforcing guidance on health and safety has driven down the numbers of injuries and deaths in the industry. Only 20 years ago over 120 construction workers died in workplace accidents each year. This is now reduced to fewer than 60 a year.

However, there is still more work to be done and it is vital that organisations such as the HSE continue to enforce health and safety and continue to reduce risks in the industry.

On-site safety inductions and toolbox talks

The HSE suggests that all new workers arriving on site should attend a short induction session on health and safety. It should:

* show the commitment of the company to health and safety

* explain the health and safety policy

* explain the roles individuals play in the policy

* state that each individual has a legal duty to contribute to safe working

* cover issues like excavations, work at height, electricity and fire risk

* provide a layout of the site and show evacuation routes

* identify where fire fighting equipment is located

* ensure that all employees have evidence of their skills

* stress the importance of signing in and out of the site.

KEY TERMS

Major injury

– any fractures, amputations, dislocations, loss of sight or other severe injury.

Over 7-day injury

– an injury that has kept someone off work for more than seven days.

DID YOU KNOW?

Workplace injuries cost the UK £13.4bn in 2010 to 2011.

Behaviour and actions that could affect others

It is the responsibility of everyone on site not only to look after their own health and safety, but also to ensure that their actions do not put anyone else at risk.

Trying to carry out work that you are not competent to do is not only dangerous to yourself but could compromise the safety of others.

Simple actions, such as ensuring that all of your rubbish and waste is properly disposed of, will go a long way to removing hazards on site that could affect others.

Just as you should not create a hazard, ignoring an obvious one is just as dangerous. You should always obey site rules and particularly the health and safety rules. You should follow any instructions you are given.

ACCIDENT AND EMERGENCY PROCEDURES

All sites will have specific procedures for dealing with accidents and emergencies. An emergency will often mean that the site needs to be evacuated, so you should know in advance where to assemble and who to report to. The site should never be re-entered without authorisation from an individual in charge or the emergency services.

Types of emergencies

Emergencies are incidents that require immediate action. They can include:

* fires
* spillages or leaks of chemicals or other hazardous substances, such as gas
* failure of a scaffold
* collapse of a wall or trench
* a health problem
* an injury
* bombs and security alerts.

Legislation and reporting accidents

RIDDOR (1995) puts a duty on employers, anyone who is self-employed, or an individual in control of the work, to report any serious workplace accidents, occupational diseases or dangerous occurrences (also known as near misses).

The report has to be made by these individuals and, if it is serious enough, the responsible person may have to fill out a RIDDOR report.

Figure 1.2 It's important that you know where your company's fire-fighting equipment is located

Injuries, diseases and dangerous occurrences

Construction sites can be dangerous places, as we have seen. The HSE maintains a list of all possible injuries, diseases and dangerous occurrences, particularly those that need to be reported.

Injuries

There are two main classifications of injuries: minor and major. A minor injury can usually be handled by a competent first aider, although it is often a good idea to refer the individual to their doctor or to the hospital. Typical minor injuries can include:

* minor cuts
* minor burns
* exposure to fumes.

Major injuries are more dangerous and will usually require the presence of an ambulance with paramedics. Major injuries can include:

* bone fracture
* concussion
* unconsciousness
* electric shock.

Diseases

There are several different diseases and health issues that have to be reported, particularly if a doctor notifies that a disease has been diagnosed. These include:

* poisoning
* infections
* skin diseases
* occupational cancer
* lung diseases
* hand/arm vibration syndrome.

Dangerous occurrences

Even if something happens that does not result in an injury, but could easily have done so, it is classed as a dangerous occurrence. It needs to be reported immediately and then followed up by an accident report form. Dangerous occurrences can include:

* accidental release of a substance that could damage health

* anything coming into contact with overhead power lines

* an electrical problem that caused a fire or explosion

* collapse or partial collapse of scaffolding over 5m high.

PRACTICAL TIP

An up-to-date list of dangerous occurrences is maintained by the Health and Safety Executive.

Recording accidents and emergencies

The Reporting of Injuries, Diseases and Dangerous Occurrences Regulations (RIDDOR) (1995) requires employers to:

* report any relevant injuries, diseases or dangerous occurrences to the Health and Safety Executive (HSE)

* keep records of incidents in a formal and organised manner (for example, in an accident book or online database).

After an accident, you may need to complete an accident report form – either in writing or online. This form may be completed by the person who was injured or the first aider.

On the accident report form you need to note down:

* the casualty's personal details, e.g. name, address, occupation
* the name of the person filling in the report form
* the details of the accident.

In addition, the person reporting the accident will need to sign the form.

On site a trained first aider will be the first individual to try and deal with the situation. In addition to trying to save life, stop the condition from getting worse and getting help, they will also record the occurrence.

On larger sites there will be a safety officer, but all businesses should keep records and documentation that details any accident or emergency that has taken place under RIDDOR and to provide that information if the HSE requests it.

Importance of reporting accidents and near misses

Reporting incidents is not just about complying with the law or providing information for statistics. Each time an accident or near miss takes place it means lessons can be learned and future problems avoided.

The accident or near miss can alert the business or organisation to a potential problem. They can then take steps to ensure that it does not occur in the future.

Major and minor injuries and near misses

RIDDOR defines a major injury as:

* a fracture (but not to a finger, thumb or toes)
* a dislocation
* an amputation
* a loss of sight in an eye
* a chemical or hot metal burn to the eye
* a penetrating injury to the eye
* an electric shock or electric burn leading to unconsciousness and/or requiring resuscitation
* hyperthermia, heat-induced illness or unconsciousness
* asphyxia
* exposure to a harmful substance
* inhalation of a substance
* acute illness after exposure to toxins or infected materials.

A minor injury could be considered as any occurrence that does not fall into any of the above categories.

A near miss is any incident that did not actually result in an injury but which could have caused a major injury if it had done so. Non-reportable near misses are useful to record as they can help to identify potential problems. Looking at a list of near misses might show patterns for potential risk.

Accident trends

We have already seen that the HSE maintains statistics on the number and types of construction accidents. The following are among the 2011/2012 construction statistics:

* There were 49 fatalities.

* There were 5,000 occupational cancer patients.

* There were 74,000 cases of work-related ill health.

* The most common types of injury were caused by falls, although many injuries were caused by falling objects, collapses and electricity. A number of construction workers were also hurt when they slipped or tripped, or were injured while lifting heavy objects.

Accidents, emergencies and the employer

Even less serious accidents and injuries can cost a business a great deal of money. But there are other costs too:

* Poor company image – if a business does not have health and safety controls in place then it may get a reputation for not caring about its employees. The number of accidents and injuries may be far higher than average.

* Loss of production – the injured individual might have to be treated and then may need a period of time off work to recover. The loss of production can include those who have to take time out from working to help the injured person and the time of a manager or supervisor who has to deal with all the paperwork and problems.

* Insurance – each time there is an accident or injury claim against the company's insurance the premiums will go up. If there are many accidents and injuries the business may find it impossible to get insurance. It is a legal requirement for a business to have insurance so in the end that company might have to close down.

* Closure of site – if there is a serious accident or injury then the site may have to be closed while investigations take place to discover the reason, or who was responsible. This could cause serious delays and loss of income for workers and the business.

DID YOU KNOW?

RoSPA (the Royal Society for the Prevention of Accidents) uses many of the statistics from the HSE. The latest figures that RoSPA has analysed date back to 2008/2009. In that year, 1.2 million people in the UK were suffering from work-related illnesses. With fewer than 132,000 reportable injuries at work, this is believed to be around half of the real figure.

DID YOU KNOW?

An employee working in a small business broke two bones in his arm. He could not return to proper duties for eight months. He lost out on wages while he was off sick and, in total, it cost the business over £45,000.

REED TIP

On some construction sites, you may get a Health and Safety Inspector come to look round without any notice – one more reason to always be thinking about working safely.

Accident and emergency authorised personnel

Several different groups of people could be involved in dealing with accident and emergency situations. These are listed in Table 1.5.

Authorised personnel	Role
First aiders and emergency responders	These are employees on site and in the workforce who have been trained to be the first to respond to accidents and injuries. The minimum provision of an appointed person would be someone who has had basic first aid training. The appointment of a first aider is someone who has attained a higher or specific level of training. A construction site with fewer than 5 employees needs an appointed first aider. A construction site with up to 50 employees requires a trained first aider, and for bigger sites at least one trained first aider is required for every 50 people.
Supervisors and managers	These have the responsibility of managing the site and would have to organise the response and contact emergency services if necessary. They would also ensure that records of any accidents are completed and up to date and notify the HSE if required.
Health and Safety Executive	The HSE requires businesses to investigate all accidents and emergencies. The HSE may send an inspector, or even a team, to investigate and take action if the law has been broken.
Emergency services	Calling the emergency services depends on the seriousness of the accident. Paramedics will take charge of the situation if there is a serious injury and if they feel it necessary will take the individual to hospital.

Table 1.5 People who deal with accident and emergency situations

DID YOU KNOW?

The three main emergency services in the UK are: the Fire Service (for fire and rescue); the Ambulance Service (for medical emergencies); the Police (for an immediate police response). Call them on 999 only if it is an emergency.

Figure 1.3 A typical first aid box

The basic first aid kit

BS 8599 relates to first aid kits, but it is not legally binding. The contents of a first aid box will depend on an employer's assessment of their likely needs. The HSE does not have to approve the contents of a first aid box but it states that where the work involves low level hazards the minimum contents of a first aid box should be:

* a copy of its leaflet on first aid – *HSE Basic advice on first aid at work*

* 20 sterile plasters of assorted size

* 2 sterile eye pads

* 4 sterile triangular bandages

* 6 safety pins

* 2 large sterile, unmedicated wound dressings

* 6 medium-sized sterile unmedicated wound dressings

* 1 pair of disposable gloves.

The HSE also recommends that no tablets or medicines are kept in the first aid box.

What to do if you discover an accident

When an accident happens it may not only injure the person involved directly, but it may also create a hazard that could then injure others. You need to make sure that the area is safe enough for you or someone else to help the injured person. It may be necessary to turn off the electrical supply or remove obstructions to the site of the accident.

The first thing that needs to be done if there is an accident is to raise the alarm. This could mean:

* calling for the first aider

* phoning for the emergency services

* dealing with the problem yourself.

How you respond will depend on the severity of the injury.

You should follow this procedure if you need to contact the emergency services:

* Find a telephone away from the emergency.

* Dial 999.

* You may have to go through a switchboard. Carefully listen to what the operator is saying to you and try to stay calm.

* When asked, give the operator your name and location, and the name of the emergency service or services you require.

* You will then be transferred to the appropriate emergency service, who will ask you questions about the accident and its location. Answer the questions in a clear and calm way.

* Once the call is over, make sure someone is available to help direct the emergency services to the location of the accident.

IDENTIFYING HAZARDS

As we have already seen, construction sites are potentially dangerous places. The most effective way of handling health and safety on a construction site is to spot the hazards and deal with them before they can cause an accident or an injury. This begins with basic housekeeping and carrying out risk assessments. It also means having a procedure in place to report hazards so that they can be dealt with.

Good housekeeping

Work areas should always be clean and tidy. Sites that are messy, strewn with materials, equipment, wires and other hazards can prove to be very dangerous. You should:

* always work in a tidy way

* never block fire exits or emergency escape routes

* never leave nails and screws scattered around

* ensure you clean and sweep up at the end of each working day

* not block walkways

* never overfill skips or bins

* never leave food waste on site.

Risk assessments and method statements

It is a legal requirement for employers to carry out risk assessments. This covers not only those who are actually working on a particular job, but other workers in the immediate area, and others who might be affected by the work.

It is important to remember that when you are carrying out work your actions may affect the safety of other people. It is important, therefore, to know whether there are any potential hazards. Once you know what these hazards are you can do something to either prevent or reduce them as a risk. Every job has potential hazards.

There are five simple steps to carrying out a risk assessment, which are shown in Table 1.6, using the example of repointing brickwork on the front face of a dwelling.

Step	Action	Example
1	Identify hazards	The property is on a street with a narrow pavement. The damaged brickwork and loose mortar need to be removed and placed in a skip below. Scaffolding has been erected. The road is not closed to traffic.
2	Identify who is at risk	The workers repointing are at risk as they are working at height. Pedestrians and vehicles passing are at risk from the positioning of the skip and the chance that debris could fall from height.
3	What is the risk from the hazard that may cause an accident?	The risk to the workers is relatively low as they have PPE and the scaffolding has been correctly erected. The risk to those passing by is higher, as they are unaware of the work being carried out above them.
4	Measures to be taken to reduce the risk	Station someone near the skip to direct pedestrians and vehicles away from the skip while the work is being carried out. Fix a secure barrier to the edge of the scaffolding to reduce the chance of debris falling down. Lower the bricks and mortar debris using a bucket or bag into the skip and not throwing them from the scaffolding. Consider carrying out the work when there are fewer pedestrians and less traffic on the road.
5	Monitor the risk	If there are problems with the first stages of the job, you need to take steps to solve them. If necessary consider taking the debris by hand through the building after removal.

Table 1.6 A five-step risk assessment for repointing brickwork

These working practices can help to prevent accidents or dangerous situations occurring in the workplace:

* *Risk assessments* look carefully at what could cause an individual harm and how to prevent this. This is to ensure that no one should be injured or become ill as a result of their work. Risk assessments identify how likely it is that an accident might happen and the consequences of it happening. A risk factor is worked out and control measures created to try to offset them.

* *Method statements,* however brief, should be available for every risk assessment. They summarise risk assessments and other findings to provide guidance on how the work should be carried out.

* *Permit to work systems* are used for very high risk or even potentially fatal activities. They are checklists that need to be completed before the work begins. They must be signed by a supervisor.

* *A hazard book* lists standard tasks and identifies common hazards. These are useful tools to help quickly identify hazards related to particular tasks.

Types of hazards

Typical construction accidents can include:

* fires and explosions

* slips, trips and falls.

* burns, including those from chemicals

* falls from scaffolding, ladders and roofs

* electrocution

* injury from faulty machinery

* power tool accidents

* being hit by construction debris

* falling through holes in flooring

We will look at some of the more common hazards in a little more detail.

Fires

Fires need oxygen, heat and fuel to burn. Even a spark can provide enough heat needed to start a fire, and anything flammable, such as petrol, paper or wood, provides the fuel. It may help to remember the 'triangle of fire' – heat, oxygen and fuel are all needed to make fire so remove one or more to help prevent or stop the fire.

Tripping

Leaving equipment and materials lying around can cause accidents, as can trailing cables and spilt water or oil. Some of these materials are also potential fire hazards.

Chemical spills

If the chemicals are not hazardous then they just need to be mopped up. But sometimes they do involve hazardous materials and there will be an existing plan on how to deal with them. A risk assessment will have been carried out.

Falls from height

A fall even from a low height can cause serious injuries. Precautions need to be taken when working at height to avoid permanent injury. You should also consider falls into open excavations as falls from height. All the same precautions need to be in place to prevent a fall.

Burns

Burns can be caused not only by fires and heat, but also from chemicals and solvents. Electricity and wet concrete and cement can also burn skin. PPE is often the best way to avoid these dangers. Sunburn is a common and uncomfortable form of burning and sunscreen should be made available. Keeping covered up, for example keeping skin covered up will help to prevent sunburn. You might think a tan looks good, but it could lead to skin cancer.

Electrical

Electricity is hazardous and electric shocks can cause burns and muscle damage, and can kill.

Exposure to hazardous substances

We look at hazardous substances in more detail on pages 20–1. COSHH regulations identify hazardous substances and require them to be labelled. You should always follow the instructions when using them.

Plant and vehicles

On busy sites there is always a danger from moving vehicles and heavy plant. Although many are fitted with reversing alarms, it may not be easy to hear them over other machinery and equipment. You should always ensure you are not blocking routes or exits. Designated walkways separate site traffic and pedestrians – this includes workers who are walking around the site. Crossing points should be in place for ease of movement on site.

Reporting hazards

We have already seen that hazards have the potential to cause serious accidents and injuries. It is therefore important to report hazards and there are different methods of doing this.

The first major reason to report hazards is to prevent danger to others, whether they are other employees or visitors to the site. It is vital to prevent accidents from taking place and to quickly correct any dangerous situations.

Injuries, diseases and actual accidents all need to be reported and so do dangerous occurrences. These are incidents that do not result in an actual injury, but could easily have hurt someone.

Accidents need to be recorded in an accident book, computer database or other secure recording system, as do near misses. Again it is a legal requirement to keep appropriate records of accidents and every company will have a procedure for this which they should tell you about. Everyone should know where the book is kept or how the records are made. Anyone that has been hurt or has taken part in dealing with an occurrence should complete the details of what has happened. Typically this will require you to fill in:

* the date, time and place of the incident

* how it happened

* what was the cause

* how it was dealt with

* who was involved

* signature and date.

The details in the book have to be transferred onto an official HSE report form.

As far as is possible, the site, company or workplace will have set procedures in place for reporting hazards and accidents. These procedures will usually be found in the place where the accident book or records are stored. The location tends to be posted on the site notice board.

How hazards are created

Construction sites are busy places. There are constantly new stages in development. As each stage is begun a whole new set of potential hazards need to be considered.

At the same time, new workers will always be joining the site. It is mandatory for them to be given health and safety instruction during induction. But sometimes this is impossible due to pressure of work or availability of trainers.

Construction sites can become even more hazardous in times of extreme weather:

* Flooding – long periods of rain can cause trenches to fill with water, cellars to be flooded and smooth surfaces to become extremely wet and slippery.

* Wind – strong winds may prevent all work at height. Scaffolding may have become unstable, unsecured roofing materials may come loose, dry-stored materials such as sand and cement may have been blown across the site.

* Heat – this can change the behaviour of materials: setting quicker, failing to cure and melting. It can also seriously affect the health of the workforce through dehydration and heat exhaustion.

* Snow – this can add enormous weight to roofs and other structures and could cause collapse. Snow can also prevent access or block exits and can mean that simple and routine work becomes impossible due to frozen conditions.

Storing combustibles and chemicals

A combustible substance can be both flammable and explosive. There are some basic suggestions from the HSE about storing these:

DID YOU KNOW?

You do not have to be involved in specialist work to come into contact with combustibles.

* Ventilation – the area should be well ventilated to disperse any vapours that could trigger off an explosion.

* Ignition – an ignition is any spark or flame that could trigger off the vapours, so materials should be stored away from any area that uses electrical equipment or any tool that heats up.

* Containment – the materials should always be kept in proper containers with lids and there should be spillage trays to prevent any leak seeping into other parts of the site.

* Exchange – in many cases it can be possible to find an alternative material that is less dangerous. This option should be taken if possible.

* Separation – always keep flammable substances away from general work areas. If possible they should be partitioned off.

Combustible materials can include a large number of commonly used substances, such as cleaning agents, paints and adhesives.

HEALTH AND HYGIENE

Just as hazards can be a major problem on site, other less obvious problems relating to health and hygiene can also be an issue. It is both your responsibility and that of your employer to make sure that you stay healthy.

KEY TERMS

Contamination

– this is when a substance has been polluted by some harmful substance or chemical.

The employer will need to provide basic welfare facilities, no matter where you are working and these must have minimum standards.

Welfare facilities

Welfare facilities can include a wide range of different considerations, as can be seen in Table 1.7.

Facilities	Purpose and minimum standards
Toilets	If there is a lock on the door there is no need to have separate male and female toilets. There should be enough for the site workforce. If there is no flushing water on site they must be chemical toilets.
Washing facilities	There should be a wash basin large enough to be able to wash up to the elbow. There should be soap, hot and cold water and, if you are working with dangerous substances, then showers are needed.
Drinking water	Clean drinking water should be available; either directly connected to the mains or bottled water. Employers must ensure that there is no contamination.
Dry room	This can operate also as a store room, which needs to be secure so that workers can leave their belongings there and also use it as a place to dry out if they have been working in wet weather, in which case a heater needs to be provided.
Work break area	This is a shelter out of the wind and rain, with a kettle, a microwave, tables and chairs. It should also have heating.

Table 1.7 Welfare facilities in the workplace

CASE STUDY

South Tyneside Homes

South Tyneside Council's Housing Company

Staying safe on site

Johnny McErlane finished his apprenticeship at South Tyneside Homes a year ago.

'I've been working on sheltered accommodation for the last year, so there are a lot of vulnerable and elderly people around. All the things I learnt at college from doing the health and safety exams comes into practice really, like taking care when using extension leads, wearing high-vis and correct footwear. It's not just about your health and safety, but looking out for others as well.

On the shelters, you can get a health and safety inspector who just comes around randomly, so you have to always be ready. It just becomes a habit once it's been drilled into you. You're health and safety conscious all the time.

The shelters also have a fire alarm drill every second Monday, so you've got to know the procedure involved there. When it comes to the more specialised skills, such as mouth-to-mouth and CPR, you might have a designated first aider on site who will have their skills refreshed regularly. Having a full first aid certificate would be valuable if you're working in construction.

You cover quite a bit of the first aid skills in college and you really have to know them because you're not always working on large sites. For example, you might be on the repairs team, working in people's houses where you wouldn't have a first aider, so you've got to have the basic knowledge yourself, just in case. All our vans have a basic first aid kit that's kept fully stocked.

The company keeps our knowledge current with these "toolbox talks", which are like refresher courses. They give you any new information that needs to be passed on to all the trades. It's a good way of keeping everyone up to date.'

Noise

Ear defenders are the best precaution to protect the ears from loud noises on site. Ear defenders are either basic ear plugs or ear muffs, which can be seen in Fig 1.13 on page 32.

The long-term impact of noise depends on the intensity and duration of the noise. Basically, the louder and longer the noise exposure, the more damage is caused. There are ways of dealing with this:

* Remove the source of the noise.

* Move the equipment away from those not directly working with it.

* Put the source of the noise into a soundproof area or cover it with soundproof material.

* Ask a supervisor if they can move all other employees away from that part of the site until the noise stops.

Substances hazardous to health

COSHH Regulations (see page 3) identify a wide variety of substances and materials that must be labelled in different ways.

Controlling the use of these substances is always difficult. Ideally, their use should be eliminated (stopped) or they should be replaced with something less harmful. Failing this, they should only be used in controlled or restricted areas. If none of this is possible then they should only be used in controlled situations.

If a hazardous situation occurs at work, then you should:

* ensure the area is made safe

* inform the supervisor, site manager, safety officer or other nominated person.

You will also need to report any potential hazards or near misses.

Personal hygiene

Construction sites can be dirty places to work. Some jobs will expose you to dust, chemicals or substances that can make contact with your skin or may stain your work clothing. It is good practice to wear suitable PPE as a first line of defence as chemicals can penetrate your skin. Whenever you have finished a job you should always wash your hands. This is certainly true before eating lunch or travelling home. It can be good practice to have dedicated work clothing, which should be washed regularly.

Always ensure you wash your hands and face and scrub your nails. This will prevent dirt, chemicals and other substances from contaminating your food and your home.

Make sure that you regularly wash your work clothing and either repair it or replace it if it becomes too worn or stained.

Health risks

The construction industry uses a wide variety of substances that could harm your health. You will also be carrying out work that could be a health risk to you, and you should always be aware that certain activities could cause long-term damage or even kill you if things go wrong. Unfortunately not all health risks are immediately obvious. It is important to make sure that from time to time you have health checks, particularly if you have been using hazardous substances. Table 1.8 outlines some potential health risks in a typical construction site.

KEY TERMS

Dermatitis

– this is an inflammation of the skin. The skin will become red and sore, particularly if you scratch the area. A GP should be consulted.

Leptospirosis

– this is also known as Weil's disease. It is spread by touching soil or water contaminated with the urine of wild animals infected with the leptospira bacteria. Symptoms are usually flu-like but in extreme cases it can cause organ failure.

Health risk	Potential future problems
Dust	The most dangerous potential dust is, of course, asbestos, which **should only be handled by specialists under controlled conditions**. But even brick dust and other fine particles can cause eye injuries, problems with breathing and even cancer.
Chemicals	Inhaling or swallowing dangerous chemicals could cause immediate, long-term damage to lungs and other internal organs. Skin problems include burns or skin can become very inflamed and sore. This is known as dermatitis.
Bacteria	Contact with waste water or soil could lead to a bacterial infection. The germs in the water or dirt could cause infection which will require treatment if they enter the body. The most extreme version is leptospirosis.
Heavy objects	Lifting heavy, bulky or awkward objects can lead to permanent back injuries that could require surgery. Heavy objects can also damage the muscles in all areas of the body.
Noise	Failure to wear ear defenders when you are exposed to loud noises can permanently affect your hearing. This could lead to deafness in the future.
Vibrating tools	Using machines that vibrate can cause a condition known as hand/arm vibration syndrome (HAVS) or vibration white finger, which is caused by injury to nerves and blood vessels. You will feel tingling that could lead to permanent numbness in the fingers and hands, as well as muscle weakness.
Cuts	Any open wound, no matter how small, leaves your body exposed to potential infections. Cuts should always be cleaned and covered, preferably with a waterproof dressing. The blood loss from deep cuts could make you feel faint and weak, which may be dangerous if you are working at height or operating machinery.
Sunlight	Most construction work involves working outside. There is a temptation to take advantage of hot weather and get a tan. But long-term exposure to sunshine means risking skin cancer so you should cover up and apply sun cream.
Head injuries	You should seek medical attention after any bump to the head. Severe head injuries could cause epilepsy, hearing problems, brain damage or death.

Table 1.8 Health risks in construction

HANDLING AND STORING MATERIALS AND EQUIPMENT

On a busy construction site it is often tempting not to even think about the potential dangers of handling equipment and materials. If something needs to be moved or collected you will just pick it up without any thought. It is also tempting just to drop your tools and other equipment when you have finished with them to deal with later. But abandoned equipment and tools can cause hazards both for you and for other people.

Safe lifting

Lifting or handling heavy or bulky items is a major cause of injuries on construction sites. So whenever you are dealing with a heavy load, it is important to carry out a basic risk assessment.

The first thing you need to do is to think about the job to be done and ask:

* Do I need to lift it manually or is there another way of getting the object to where I need it?

Consider any mechanical methods of transporting loads or picking up materials. If there really is no alternative, then ask yourself:

1. Do I need to bend or twist?
2. Does the object need to be lifted or put down from high up?
3. Does the object need to be carried a long way?
4. Does the object need to be pushed or pulled for a long distance?
5. Is the object likely to shift around while it is being moved?

If the answer to any of these questions is 'yes', you may need to adjust the way the task is done to make it safer.

Think about the object itself. Ask:

1. Is it just heavy or is it also bulky and an awkward shape?
2. How easy is it to get a good hand-hold on the object?
3. Is the object a single item or are there parts that might move around and shift the weight?
4. Is the object hot or does it have sharp edges?

Again, if you have answered 'yes' to any of these questions, then you need to take steps to address these issues.

It is also important to think about the working environment and where the lifting and carrying is taking place. Ask yourself:

1. Are the floors stable?

2. Are the surfaces slippery?

3. Will a lack of space restrict my movement?

4. Are there any steps or slopes?

5. What is the lighting like?

Before lifting and moving an object, think about the following:

* Check that your pathway is clear to where the load needs to be taken.

* Look at the product data sheet and assess the weight. If you think the object is too heavy or difficult to move then ask someone to help you. Alternatively, you may need to use a mechanical lifting device.

When you are ready to lift, gently raise the load. Take care to ensure the correct posture – you should have a straight back, with your elbows tucked in, your knees bent and your feet slightly apart.

Once you have picked up the load, move slowly towards your destination. When you get there, make sure that you do not drop the load but carefully place it down.

DID YOU KNOW?

Although many people regard the weight limit for lifting and/or moving heavy or awkward objects to be 20 kg, the HSE does not recommend safe weights. There are many things that will affect the ability of an individual to lift and carry particular objects and the risk that this creates, so manual handling should be avoided altogether where possible.

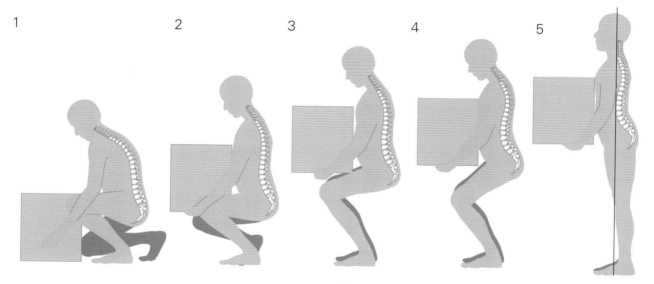

Figure 1.4 Take care to follow the correct procedure for lifting

Sack trolleys are useful for moving heavy and bulky items around. Gently slide the bottom of the sack trolley under the object and then raise the trolley to an angle of 45° before moving off. Make sure that the object is properly balanced and is not too big for the trolley.

Trailers and forklift trucks are often used on large construction sites, as are dump trucks. Never use these without proper training.

Figure 1.5 Pallet truck

Figure 1.6 Sack trolley

Site safety equipment

You should always read the construction site safety rules and when required wear your PPE. Simple things, such as wearing the right footwear for the right job, are important.

Safety equipment falls into two main categories:

* PPE – including hard hats, footwear, gloves, glasses and safety vests

* perimeter safety – this includes screens, netting and guards or clamps to prevent materials from falling or spreading.

Construction safety is also directed by signs, which will highlight potential hazards.

Safe handling of materials and equipment

All tools and equipment are potentially dangerous. It is up to you to make sure that they do not cause harm to yourself or others. You should always know how to use tools and equipment. This means either instruction from someone else who is experienced, or at least reading the manufacturer's instructions.

You should always make sure that you:

* use the right tool – don't be tempted to use a tool that is close to hand instead of the one that is right for the job

* wear your PPE – the one time you decide not to bother could be the time that you injure yourself

* never try to use a tool or a piece of equipment that you have not been trained to use.

You should always remember that if you are working on a building that was constructed before 2000 it may contain asbestos.

Correct storage

We have already seen that tools and equipment need to be treated with respect. Damaged tools and equipment are not only less effective at doing their job, they could also cause you to injure yourself.

Table 1.9 provides some pointers on how to store and handle different types of materials and equipment.

Materials and equipment	Safe storage and handling
Hand tools	Store hand tools with sharp edges either in a cover or a roll. They should be stored in bags or boxes. They should always be dried before putting them away as they will rust.
Power tools	Never carry them by the cable. Store them in their original carrying case. Always follow the manufacturer's instructions.
Wheelbarrows	Check the tyres and metal stays regularly. Always clean out after use and never overload.
Bricks and blocks	Never store more than two packs high. When cutting open a pack, be careful as the bricks could collapse.
Slabs and curbs	Store slabs flat on their edges on level ground, preferably with wood underneath to prevent damage. Store curbs the same way. To prevent weather damage, cover them with a sheet.
Tiles	Always cover them and protect them from damage as they are relatively fragile. Ideally store them in a hut or container.
Aggregates	Never store aggregates under trees as leaves will drop on them and contaminate them. Cover them with plastic sheets.
Plaster and plasterboard	Plaster needs to be kept dry, so even if stored inside you should take the precaution of putting the bags on pallets. To prevent moisture do not store against walls and do not pile higher than five bags. Plasterboard can be awkward to manage and move around. It also needs to be stored in a waterproof area. It should be stored flat and off the ground but should not be stored against walls as it may bend. Use a rotation system so that the materials are not stored in the same place for long periods.
Wood	Always keep wood in dry, well-ventilated conditions. If it needs to be stored outside it should be stored on bearers that may be on concrete. If wood gets wet and bends it is virtually useless. Always be careful when moving large cuts of wood or sheets of ply or MDF as they can easily become damaged.
Adhesives and paint	Always read the manufacturer's instructions. Ideally they should always be stored on clearly marked shelves. Make sure you rotate the stock using the older stock first. Always make sure that containers are tightly sealed. Storage areas must comply with fire regulations and display signs to advise of their contents.

Table 1.9 Safe storing and handling of materials and equipment

Waste control

The expectation within the building services industry is increasingly that working practices conserve energy and protect the environment. Everyone can play a part in this. For example, you can contribute by turning off hose pipes when you have finished using water, or not running electrical items when you don't need to.

Simple things, such as keeping construction sites neat and orderly, can go a long way to conserving energy and protecting the environment. A good way to remember this is Sort, Set, Shine, Standardise:

* Sort – sort and store items in your work area, eliminate clutter and manage deliveries.

* Set – everything should have its own place and be clearly marked and easy to access. In other words, be neat!

Figure 1.7 It's important to create as little waste as possible on the construction site

* Shine – clean your work area and you will be able to see potential problems far more easily.

* Standardise – by using standardised working practices you can keep organised, clean and safe.

Reducing waste is all about good working practice. By reducing wastage disposal, and recycling materials on site, you will benefit from savings on raw materials and lower transportation costs.

Planning ahead, and accurately measuring and cutting materials, means that you will be able to reduce wastage.

BASIC WORKING PLATFORMS AND ACCESS EQUIPMENT

Working at height should be eliminated or the work carried out using other methods where possible. However, there may be situations where you may need to work at height. These situations can include:

* roofing

* repair and maintenance above ground level

* working on high ceilings.

Any work at height must be carefully planned. Access equipment includes all types of ladder, scaffold and platform. You must always use a working platform that is safe. Sometimes a simple step ladder will be sufficient, but at other times you may have to use a tower scaffold.

Generally, ladders are fine for small, quick jobs of less than 30 minutes. However, for larger, longer jobs a more permanent piece of access equipment will be necessary.

Working platforms and access equipment: good practice and dangers of working at height

Table 1.10 outlines the common types of equipment used to allow you to work at heights, along with the basic safety checks necessary.

Equipment	Main features	Safety checks
Step ladder	Ideal for confined spaces. Four legs give stability	• Knee should remain below top of steps • Check hinges, cords or ropes • Position only to face work
Ladder	Ideal for basic access, short-term work. Made from aluminium, fibreglass or wood	• Check rungs, tie rods, repairs, and ropes and cords on stepladders • Ensure it is placed on firm, level ground • Angle should be no greater than 75° or 1 in 4
Mobile mini towers or scaffolds	These are usually aluminium and foldable, with lockable wheels	• Ensure the ground is even and the wheels are locked • Never move the platform while it has tools, equipment or people on it
Roof ladders and crawling boards	The roof ladder allows access while crawling boards provide a safe passage over tiles	• The ladder needs to be long enough and supported • Check boards are in good condition • Check the welds are intact • Ensure all clips function correctly
Mobile tower scaffolds	These larger versions of mini towers usually have edge protection	• Ensure the ground is even and the wheels are locked • Never move the platform while it has tools, equipment or people on it • Base width to height ratio should be no greater than 1:3
Fixed scaffolds and edge protection	Scaffolds fitted and sized to the specific job, with edge protection and guard rails	• There needs to be sufficient braces, guard rails and scaffold boards • The tubes should be level • There should be proper access using a ladder
Mobile elevated work platforms	Known as scissor lifts or cherry pickers	• Specialist training is required before use • Use guard rails and toe boards • Care needs to be taken to avoid overhead hazards such as cables

Table 1.10 Equipment for working at height and safety checks

You must be trained in the use of certain types of access equipment, like mobile scaffolds. Care needs to be taken when assembling and using access equipment. These are all examples of good practice:

* Step ladders should always rest firmly on the ground. Only use the top step if the ladder is part of a platform.

* Do not rest ladders against fragile surfaces, and always use both hands to climb. It is best if the ladder is steadied (footed) by someone at the foot of the ladder. Always maintain three points of contact – two feet and one hand.

* A roof ladder is positioned by turning it on its wheels and pushing it up the roof. It then hooks over the ridge tiles. Ensure that the access ladder to the roof is directly beside the roof ladder.

* A mobile scaffold is put together by slotting sections until the required height is reached. The working platform needs to have a suitable edge protection such as guard-rails and toe-boards. Always push from the bottom of the base and not from the top to move it, otherwise it may lean or topple over.

Figure 1.8 A tower scaffold

WORKING SAFELY WITH ELECTRICITY

It is essential whenever you work with electricity that you are competent and that you understand the common dangers. Electrical tools must be used in a safe manner on site. There are precautions that you can take to prevent possible injury, or even death.

Precautions

Whether you are using electrical tools or equipment on site, you should always remember the following:

* Use the right tool for the job.

* Use a transformer with equipment that runs on 110V.

* Keep the two voltages separate from each other. You should avoid using 230V where possible but use a residual current device (RCD) if you have to use 230V.

* When using 100V, ensure that leads are yellow in colour.

* Check the plug is in good order

* Confirm that the fuse is the correct rating for the equipment.

* Check the cable (including making sure that it does not present a tripping hazard).

* Find out where the mains switch is, in case you need to turn off the power in the event of an emergency.

* Never attempt to repair electrical equipment yourself.

* Disconnect from the mains power before making adjustments, such as changing a drill bit.

* Make sure that the electrical equipment has a sticker that displays a recent test date.

Visual inspection and testing is a three-stage process:

1. The user should check for potential danger signs, such as a frayed cable or cracked plug.

2. A formal visual inspection should then take place. If this is done correctly then most faults can be detected.

3. Combined inspections and PAT should take place at regular intervals by a competent person.

Watch out for the following causes of accidents – they would also fail a safety check:

KEY TERMS

PAT

– Portable Appliance Testing – regular testing is a health and safety requirement under the Electricity at Work Regulations (1989).

- damage to the power cable or plug
- taped joints on the cable
- wet or rusty tools and equipment
- weak external casing
- loose parts or screws

- signs of overheating
- the incorrect fuse
- lack of cord grip
- electrical wires attached to incorrect terminals
- bare wires.

DID YOU KNOW?

All power tools should be checked by the user before use. A PAT programme of maintenance, inspection and testing is necessary. The frequency of inspection and testing will depend on the appliance. Equipment is usually used for a maximum of three months between tests.

When preparing to work on an electrical circuit, do not start until a permit to work has been issued by a supervisor or manager to a competent person.

Make sure the circuit is broken before you begin. A 'dead' circuit will not cause you, or anybody else, harm. These steps must be followed:

- Switch off – ensure the supply to the circuit is switched off by disconnecting the supply cables or using an isolating switch.
- Isolate – disconnect the power cables or use an isolating switch.
- Warn others – to avoid someone reconnecting the circuit, place warning signs at the isolation point.
- Lock off – this step physically prevents others from reconnecting the circuit.
- Testing – is carried out by electricians but you should be aware that it involves three parts:
  1. testing a voltmeter on a known good source (a live circuit) so you know it is working properly
  2. checking that the circuit to be worked on is dead
  3. rechecking your voltmeter on the known live source, to prove that it is still working properly.

It is important to make sure that the correct point of isolation is identified. Isolation can be next to a local isolation device, such as a plug or socket, or a circuit breaker or fuse.

The isolation should be locked off using a unique key or combination. This will prevent access to a main isolator until the work has been completed. Alternatively, the handle can be made detachable in the OFF position so that it can be physically removed once the circuit is switched off.

Dangers

You are likely to encounter a number of potential dangers when working with electricity on construction sites or in private houses. Table 1.11 outlines the most common dangers.

Danger	Identifying the danger
Faulty electrical equipment	Visually inspect for signs of damage. Equipment should be double insulated or incorporate an earth cable.
Damaged or worn cables	Check for signs of wear or damage regularly. This includes checking power tools and any wiring in the property.
Trailing cables	Cables lying on the ground, or worse, stretched too far, can present a tripping hazard. They could also be cut or damaged easily.
Cables and pipe work	Always treat services you find as though they are live. This is very important as services can be mistaken for one another. You may have been trained to use a cable and pipe locator that finds cables and metal pipes.
Buried or hidden cables	Make sure you have plans. Alternatively, use a cable and pipe locator, mark the positions, look out for signs of service connection cables or pipes and hand-dig trial holes to confirm positions.
Inadequate over-current protection	Check circuit breakers and fuses are the correct size current rating for the circuit. A qualified electrician may have to identify and label these.

Table 1.11 Common dangers when working with electricity

Each year there are around 1,000 accidents at work involving electric shocks or burns from electricity. If you are working in a construction site you are part of a group that is most at risk. Electrical accidents happen when you are working close to equipment that you think is disconnected but which is, in fact, live.

Another major danger is when electrical equipment is either misused or is faulty. Electricity can cause fires and contact with the live parts can give you an electric shock or burn you.

Different voltages

The two most common voltages that are used in the UK are 230 V and 110 V:

* 230 V: this is the standard domestic voltage. But on construction sites it is considered to be unsafe and therefore 110 V is commonly used.

* 110 V: these plugs are marked with a yellow casement and they have a different shaped plug. A transformer is required to convert 230 V to 110 V.

Some larger homes, as well as industrial and commercial buildings, may have 415 V supplies. This is the same voltage that is found on overhead electricity cables. In most houses and other buildings the voltage from these cables is reduced to 230 V. This is what most electrical equipment works from. Some larger machinery actually needs 415 V.

In these buildings the 415 V comes into the building and then can either be used directly or it is reduced so that normal 230 V appliances can be used.

Colour coded cables

Normally you will come across three differently coloured wires: Live, Neutral and Earth. These have standard colours that comply with European safety standards and to ensure that they are easily identifiable. However, in some older buildings the colours are different.

Wire type	Modern colour	Older colour
Live	Brown	Red
Neutral	Blue	Black
Earth	Yellow and Green	Yellow and Green

Table 1.12 Colour coding of cables

Working with equipment with different electrical voltages

You should always check that the electrical equipment that you are going to use is suitable for the available electrical supply. The equipment's power requirements are shown on its rating plate. The voltage from the supply needs to match the voltage that is required by the equipment.

Storing electrical equipment

Electrical equipment should be stored in dry and secure conditions. Electrical equipment should never get wet but – if it does happen – it should be dried before storage. You should always clean and adjust the equipment before connecting it to the electricity supply.

PERSONAL PROTECTIVE EQUIPMENT (PPE)

Personal protective equipment, or PPE, is a general term that is used to describe a variety of different types of clothing and equipment that aim to help protect against injuries or accidents. Some PPE you will use on a daily basis and others you may use from time to time. The type of PPE you wear depends on what you are doing and where you are. For example, the practical exercises in this book were photographed at a college, which has rules and requirements for PPE that are different to those on large construction sites. Follow your tutor's or employer's instructions at all times.

Types of PPE

PPE literally covers from head to foot. Here are the main PPE types.

Figure 1.9 A hi-vis jacket

Figure 1.10 Safety glasses and goggles

Figure 1.11 Hand protection

Figure 1.12 Head protection

Figure 1.13 Hearing protection

Protective clothing

Clothing protection such as overalls:

* provides some protection from spills, dust and irritants
* can help protect you from minor cuts and abrasions
* reduces wear to work clothing underneath.

Sometimes you may need waterproof or chemical-resistant overalls.

High visibility (hi-vis) clothing stands out against any background or in any weather conditions. It is important to wear high visibility clothing on a construction site to ensure that people can see you easily. In addition, workers should always try to wear light-coloured clothing underneath, as it is easier to see.

You need to keep your high visibility and protective clothing clean and in good condition.

Employers need to make sure that employees understand the reasons for wearing high visibility clothing and the consequences of not doing so.

Eye protection

For many jobs, it is essential to wear goggles or safety glasses to prevent small objects, such as dust, wood or metal, from getting into the eyes. As goggles tend to steam up, particularly if they are being worn with a mask, safety glasses can often be a good alternative.

Hand protection

Wearing gloves will help to prevent damage or injury to the hands or fingers. For example, general purpose gloves can prevent cuts, and rubber gloves can prevent skin irritation and inflammation, such as contact dermatitis caused by handling hazardous substances. There are many different types of gloves available, including specialist gloves for working with chemicals.

Head protection

Hard hats or safety helmets are compulsory on building sites. They can protect you from falling objects or banging your head. They need to fit well and they should be regularly inspected and checked for cracks. Worn straps mean that the helmet should be replaced, as a blow to the head can be fatal. Hard hats bear a date of manufacture and should be replaced after about 3 years.

Hearing protection

Ear defenders, such as ear protectors or plugs, aim to prevent damage to your hearing or hearing loss when you are working with loud tools or are involved in a very noisy job.

Respiratory protection

Breathing in fibre, dust or some gases could damage the lungs. Dust is a very common danger, so a dust mask, face mask or respirator may be necessary.

Make sure you have the right mask for the job. It needs to fit properly otherwise it will not give you sufficient protection.

Foot protection

Foot protection is compulsory on site. Footwear should include steel toecaps (or equivalent) to protect feet from dropped objects, midsole protection (usually a steel plate) to protect against puncture or penetration from things like nails on the floor, and soles with good grip to help prevent slips on wet surfaces.

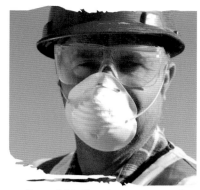

Figure 1.14 Respiratory protection

Legislation covering PPE

The most important piece of legislation is the Personal Protective Equipment at Work Regulations (1992). It covers all sorts of PPE and sets out your responsibilities and those of the employer. Linked to this are the Control of Substances Hazardous to Health (2002) and the Provision and Use of Work Equipment Regulations (1992 and 1998).

Storing and maintaining PPE

All forms of PPE will be less effective if they are not properly maintained. This may mean examining the PPE and either replacing or cleaning it, or if relevant testing or repairing it. PPE needs to be stored properly so that it is not damaged, contaminated or lost. Each type of PPE should have a CE mark. This shows that it has met the necessary safety requirements.

Importance of PPE

PPE needs to be suitable for its intended use and it needs to be used in the correct way. As a worker or an employee you need to:

* make sure you are trained to use PPE

* follow your employer's instructions when using the PPE and always wear it when you are told to do so

* look after the PPE and if there is a problem with it report it.

Your employer will:

* know the risks that the PPE will either reduce or avoid

* know how the PPE should be maintained

* know its limitations.

Consequences of not using PPE

The consequences of not using PPE can be immediate or long-term. Immediate problems are more obvious, as you may injure yourself. The longer-term consequences could be ill health in the future. If your employer has provided PPE, you have a legal responsibility to wear it.

FIRE AND EMERGENCY PROCEDURES

KEY TERMS

Assembly point

– an agreed place outside the building to go to if there is an emergency.

If there is a fire or an emergency, it is vital that you raise the alarm quickly. You should leave the building or site and then head for the **assembly point.**

When there is an emergency a general alarm should sound. If you are working on a larger and more complex construction site, evacuation may begin by evacuating the area closest to the emergency. Areas will then be evacuated one-by-one to avoid congestion of the escape routes.

Three elements essential to creating a fire

Three ingredients are needed to make something combust (burn):

* oxygen   * heat   * fuel.

The fuel can be anything which burns, such as wood, paper or flammable liquids or gases, and oxygen is in the air around us, so all that is needed is sufficient heat to start a fire.

The fire triangle represents these three elements visually. By removing one of the three elements the fire can be prevented or extinguished.

Figure 1.15 Assembly point sign

How fire is spread

Fire can easily move from one area to another by finding more fuel. You need to consider this when you are storing or using materials on site, and be aware that untidiness can be a fire risk. For example, if there are wood shavings on the ground the fire can move across them, burning up the shavings.

Figure 1.16 The fire triangle

Heat can also transfer from one source of fuel to another. If a piece of wood is on fire and is against or close to another piece of wood, that too will catch fire and the fire will have spread.

On site, fires are classified according to the type of material that is on fire. This will determine the type of fire-fighting equipment you will need to use. The five different types of fire are shown in Table 1.13.

Class of fire	Fuel or material on fire
A	Wood, paper and textiles
B	Petrol, oil and other flammable liquids
C	LPG, propane and other flammable gases
D	Metals and metal powder
E	Electrical equipment

Table 1.13 Different classes of fire

There is also F, cooking oil, but this is less likely to be found on site, except in a kitchen.

Taking action if you discover a fire and fire evacuation procedures

During induction, you will have been shown what to do in the event of a fire and told about assembly points. These are marked by signs and somewhere on the site there will be a map showing their location.

If you discover a fire you should:

* sound the alarm

* not attempt to fight the fire unless you have had fire marshal training

* otherwise stop work, do not collect your belongings, do not run, and do not re-enter the site until the all clear has been given.

Different types of fire extinguishers

Extinguishers can be effective when tackling small localised fires. However, you must use the correct type of extinguisher. For example, putting water on an oil fire could make it explode. For this reason, you should not attempt to use a fire extinguisher unless you have had proper training.

When using an extinguisher it is important to remember the following safety points:

* Only use an extinguisher at the early stages of a fire, when it is small.

* The instructions for use appear on the extinguisher.

* If you do choose to fight the fire because it is small enough, and you are sure you know what is burning, position yourself between the fire and the exit, so that if it doesn't work you can still get out.

Type of fire risk	Fire class Symbol	White label Water	Cream label Foam	Black label Carbon dioxide	Blue label Dry powder	Yellow label Wet chemical
A – Solid (e.g. wood or paper)	A	✓	✓	✗	✓	✓
B – Liquid (e.g. petrol)	B	✗	✓	✓	✓	✗
C – Gas (e.g. propane)	C	✗	✗	✓	✓	✗
D – Metal (e.g. aluminium)	D METAL	✗	✗	✗	✓	✗
E – Electrical (i.e. any electrical equipment)	E	✗	✗	✓	✓	✗
F – Cooking oil (e.g. a chip pan)	F	✗	✗	✗	✗	✓

Table 1.14 Types of fire extinguishers

There are some differences you should be aware of when using different types of extinguisher:

* CO_2 *extinguishers* – do not touch the nozzle; simply operate by holding the handle. This is because the nozzle gets extremely cold when ejecting the CO_2, as does the canister. Fires put out with a CO_2 extinguisher may reignite, and you will need to ventilate the room after use.

* *Powder extinguishers* – these can be used on lots of kinds of fire, but can seriously reduce visibility by throwing powder into the air as well as on the fire.

SIGNS AND SAFETY NOTICES

In a well-organised working environment safety signs will warn you of potential dangers and tell you what to do to stay safe. They are used to warn you of hazards. Their purpose is to prevent accidents. Some will tell you what to do (or not to do) in particular parts of the site and some will show you where things are, such as the location of a first aid box or a fire exit.

Types of signs and safety notices

There are five basic types of safety sign, as well as signs that are a combination of two or more of these types. These are shown in Table 1.15.

Type of safety sign	What it tells you	What it looks like	Example
Prohibition sign	Tells you what you must *not* do	Usually round, in red and white	Do not use ladder
Hazard sign	Warns you about hazards	Triangular, in yellow and black	Caution Slippery floor
Mandatory sign	Tells you what you *must* do	Round, usually blue and white	Masks must be worn in this area
Safe condition or information sign	Gives important information, e.g. about where to find fire exits, assembly points or first aid kit, or about safe working practices	Green and white	First aid
Firefighting sign	Gives information about extinguishers, hydrants, hoses and fire alarm call points, etc.	Red with white lettering	Fire alarm call point
Combination sign	These have two or more of the elements of the other types of sign, e.g. hazard, prohibition and mandatory		DANGER Isolate before removing cover

Table 1.15 Different types of safety signs

TEST YOURSELF

1. Which of the following requires you to tell the HSE about any injuries or diseases?

   a. HASAWA

   b. COSHH

   c. RIDDOR

   d. PUWER

2. What is a prohibition notice?

   a. An instruction from the HSE to stop all work until a problem is dealt with

   b. A manufacturer's announcement to stop all work using faulty equipment

   c. A site contractor's decision not to use particular materials

   d. A local authority banning the use of a particular type of brick

3. Which of the following is considered a major injury?

   a. Bruising on the knee

   b. Cut

   c. Concussion

   d. Exposure to fumes

4. If there is an accident on a site who is likely to be the first to respond?

   a. First aider

   b. Police

   c. Paramedics

   d. HSE

5. Which of the following is a summary of risk assessments and is used for high risk activities?

   a. Permit to work

   b. Hazard book

   c. Monitoring statement

   d. Method statement

6. Some substances are combustible. Which of the following are examples of combustible materials?

   a. Adhesives

   b. Paints

   c. Cleaning agents

   d. All of these

7. What is dermatitis?

   a. Inflammation of the skin

   b. Inflammation of the ear

   c. Inflammation of the eye

   d. Inflammation of the nose

8. Screens, netting and guards on a site are all examples of which of the following?

   a. PPE

   b. Signs

   c. Perimeter safety

   d. Electrical equipment

9. Which of the following are also known as scissor lifts or cherry pickers?

   a. Bench saws

   b. Hand-held power tools

   c. Cement additives

   d. Mobile elevated work platforms

10. In older properties the neutral electricity wire is which colour?

    a. Black

    b. Red

    c. Blue

    d. Brown

Chapter 2

Unit CSA–L1Core02

KNOWLEDGE OF TECHNICAL INFORMATION, QUANTITIES AND COMMUNICATION WITH OTHERS

LEARNING OUTCOMES

LO1: Know how to interpret construction related technical information

LO2: Know how to determine quantities of materials

LO3: Know how to relay information in the construction environment

LO4: Know how to communicate with others in the construction environment

INTRODUCTION

The aim of this chapter is to:

* show you the processes of passing on information
* show you the concepts of effective communication.

INTERPRETING CONSTRUCTION-RELATED TECHNICAL INFORMATION

Even quite simple construction projects will require documents. These provide you with the necessary information you will need to do the job. The documents are produced by a range of different people and each document has a different purpose. Together they give you the full picture of the job, from the basic outline through to the technical specifications.

Importance of documentation

In many industries a great deal of information is only ever stored electronically. This is not always an option in the construction industry. Many documents, such as working drawings, will need to be referred to on site. Detailed drawings of components that need to be made will have to be measured and checked before making joints, for example, in the workshop.

It is not always easy to store and look after working documents. The following advice is worth remembering:

* Always ensure you have the latest version of a document to work from before you begin to follow its instructions.

* If you are not going to need to use a document until later then get into the habit of storing it somewhere safe.

* Try to make sure that you do not leave documents lying around on site, where they could get lost or damaged.

* Try to make sure you always have a second copy of the document. You should keep this away from the site, in reserve, in case you lose your working copy.

* You should store any documents that you have used on a particular job at least until that job is completely finished.

* You might need to store the documents for some time after in case you need to refer back to them for repair and servicing.

Interpreting construction specifications

Obviously it would be impossible to put in all of the details in full, so symbols, hatchings and abbreviations are used to simplify the drawings. All of these symbols or hatchings are drawn to follow a British Standards-approved format, BS 1192. The symbols cover various types of brickwork and blockwork, as well as concrete, hard core and insulation, as can be seen in Fig 2.1.

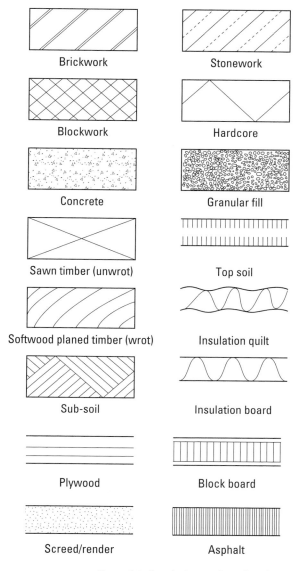

Figure 2.1 Symbols used on drawings

Common abbreviations

For the same reason, abbreviations are often used. Table 2.1 outlines some examples that you will need to become familiar with.

Abbreviation	Meaning
bwk	Brickwork
conc	Areas that will be concreted
dpc	Damp-proof course
fdn	Foundations
insul	Insulation
rwg	Rainwater gulleys
svp	Soil and vent pipe

Table 2.1

Types of documentation

Supporting information can be found in a variety of different types of documents. These include:

* drawings and plans
* programmes of work
* procedures
* specifications
* policies
* schedules
* manufacturers' technical information
* organisational documentation
* training and development records
* risk and method statements
* Construction (Design and Management) (CDM) Regulations
* Building Regulations.

Drawings and plans

Drawings are an important part of construction work. You will need to understand how drawings provide you with the information you need to carry out the work. The drawings show what the building will look like and how it will be constructed. This means that there are several different drawings of the building from different viewpoints.

Block plans

Block plans show the construction site and the surrounding area. Normally block plans are at a ratio of 1:2500 (usually in rural areas) and 1:1250 (usually in urban areas). This means that 1 mm on a block plan is equal to 2,500 mm or 1,250 mm on the ground.

Site plan

The site plan drawing shows what is basically planned for the site. It is an important drawing because it has been created in order to get approval for the project from planning committees or funding sources. In most cases the site plan is an architectural plan, showing the basic arrangement of buildings and any landscaping.

The site plan will usually show:

* directional orientation (i.e. the north point)

* location and size of the building or buildings

* existing structures

* clear measurements.

General location

Location drawings show the site or building in relation to its surroundings. It will therefore show details such as boundaries, other buildings and roads. It will also contain other vital information, including:

* access
* sewers

* drainage
* the north point.

The scale will also be shown and the drawing will have a title. It will also be given a job or project number to help identify it easily, as well as an address, the date of the drawing and the name of the client. A version number will also be on the drawing with an amendment date if there have been any changes. You'll need to make sure you have the latest drawing.

Normally location drawings are either 1:500 or 1:200 (that is, 1 mm of the drawing represents 500 mm or 200 mm on the ground).

Assembly

These are detailed drawings that illustrate the different elements and components of the construction. They tend to be 1:20, 1:10 or 1:5 (1 cm of the drawing represents 20 mm, 10 mm or 5 mm on the ground). This larger scale allows more detail to be shown, to ensure accurate construction.

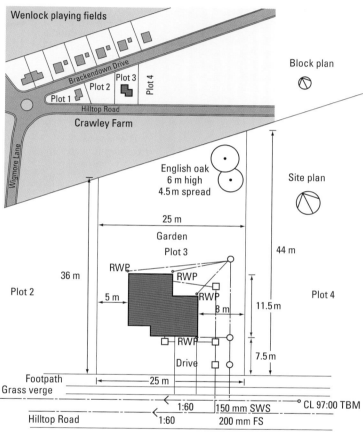

Figure 2.2 Block plan

Figure 2.3 Location plan

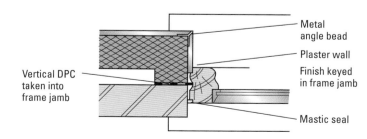

Figure 2.4 Assembly drawing

Sectional

These drawings aim to provide:

* vertical dimensions

* horizontal dimensions

* constructional details.

They can be used to show the height of ground levels, damp-proof courses, foundations and other aspects of the construction.

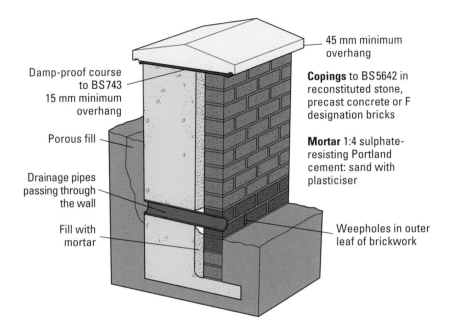

Damp-proof course to BS 743 15 mm minimum overhang

Porous fill

Drainage pipes passing through the wall

Fill with mortar

45 mm minimum overhang

Copings to BS 5642 in reconstituted stone, precast concrete or F designation bricks

Mortar 1:4 sulphate-resisting Portland cement: sand with plasticiser

Weepholes in outer leaf of brickwork

Figure 2.5 Section drawing of an earth retaining wall

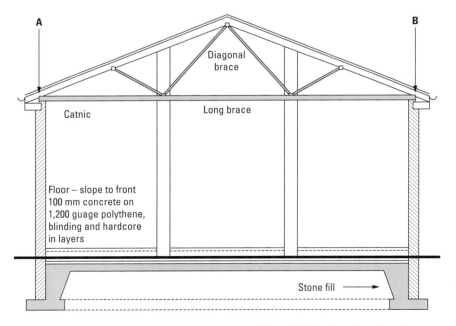

A

B

Diagonal brace

Catnic

Long brace

Floor – slope to front 100 mm concrete on 1,200 guage polythene, blinding and hardcore in layers

Stone fill →

Figure 2.6 Section drawing of a garage

Detail drawings

These drawings show how a component needs to be manufactured. They are used to show the relationship between different components within the fabric of the building. For instance, an eaves detail would show rafters, wall plate, roof coverings, inner and outer masonry, insulation and much more. Details can be shown in various scales, but mainly 1:10, 1:5 and 1:1 (the same size as the actual component if it is small).

Orthographic projection (first angle)

First angle projection is a view that represents the side view, the front view and the plan view from above, as can be seen in Fig 2.8.

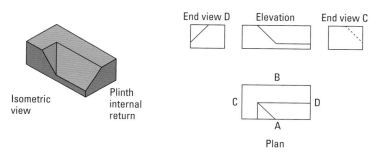

Figure 2.7 First angle projection

Isometric projection

Isometric projection is a way of representing three dimensional objects in two dimensions, as can also be seen in Fig 2.8. All horizontal lines are drawn at 30°.

Programmes of work

Programmes of work show the actual sequence of any work activities on a construction project. Part of the work programme plan is to show target times. They are usually shown in the form of a bar or Gantt chart (a special kind of bar chart), as can be seen in Fig 2.9.

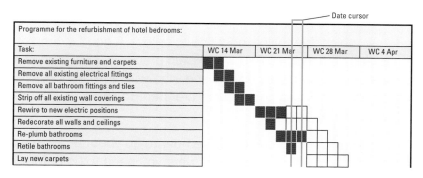

Figure 2.9 Single line contract plan Gantt chart

Serving hatch   Vertical section

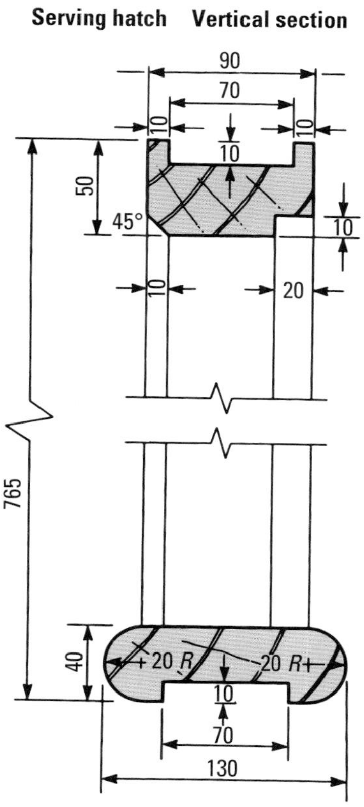

Figure 2.8 Detail drawing (measurements in mm)

DID YOU KNOW?

First angle is also known as European projection because Americans use third angle projection, which shows the views from a different position.

This figure shows the following:

* On the left hand side all of the tasks are listed – note this is ordered in the sequence of construction.

* On the right the blocks show the target start and end date for each of the individual tasks.

* The timescale can be either days, weeks or months.

Far more complex forms of work programmes can also be created. Fig 2.10 shows the construction of a house.

This more complex example of a Gantt chart shows the following:

* There are two lines – they show the target dates and actual dates. The actual dates are shaded, showing when the work actually began and how long it actually took.

* If this Gantt chart is kept up to date an accurate picture of progress and estimated completion time can be seen.

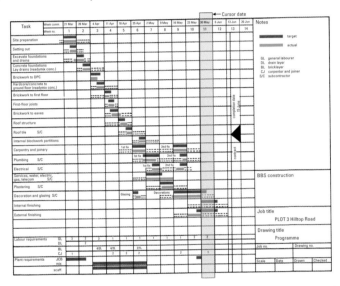

Figure 2.10 Gantt chart for the construction of a house

Procedures

When you work for a construction company there will be a series of procedures, which they will expect you to follow. A good example is the emergency procedure. This will explain precisely what is required in the case of an emergency on site and who will have responsibility to carry out particular duties. Procedures are there to show you the right way of doing something.

A construction procedure could outline how to go about building a wall or hanging a door, taking into account what you need to do beforehand, the materials and tools that are required, and the order in which you must carry out each step.

Another good example of a procedure is the procurement or buying procedure. This will outline:

* who is authorised to buy what, and how much individuals are allowed to spend

* any forms or documents that have to be completed when buying.

Specifications

In addition to drawings it is usually necessary to have documents known as specifications. These provide much more information, as can be seen in Fig 2.11.

The specifications give you a precise description. They will include:

* the address and description of the site

* on-site services (e.g. water and electricity)

* materials description, outlining the size, finish, quality and tolerances

* specific requirements, such as the individual who will authorise or approve work carried out

* any restrictions on site, such as working hours.

Policies

Policies are sets of principles or a programme of actions. The following are two good examples:

* The environmental policy outlines how the business goes about protecting the environment.

* The safety policy outlines how the business deals with health and safety matters and who is responsible for monitoring and maintaining it.

You will normally find both policies and procedures in site rules. These are usually explained to each new employee when they first join the company. Sometimes there may be additional site rules, depending on the job and the location of the work.

Schedules

Schedules are cross-referenced to drawings that have been prepared by an architect. They will show specific design information. Usually they are prepared for jobs that will be carried out regularly on site, such as:

* working on windows, doors, floors, walls or ceilings

* working on drainage, lintels or sanitary ware.

A schedule can be seen in Fig 2.12.

The schedule is very useful for:

* working out the quantities of materials needed

* ordering materials and components and then checking them against deliveries

* locating where specific materials will be used.

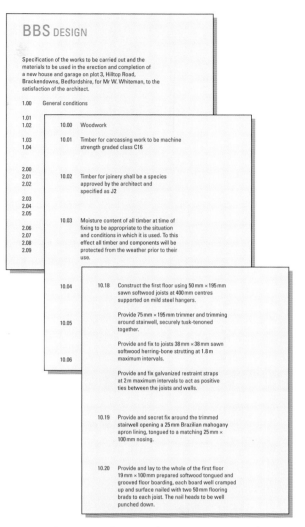

Figure 2.11 Extracts from a typical specification

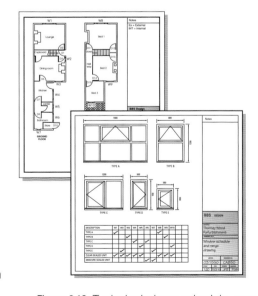

Figure 2.12 Typical windows schedule, range drawing and floor plans

Manufacturers' technical information

Almost everything that is bought to be used on site will come with a variety of information. The basic technical information provided will show what the equipment or material is intended to be used for, how it should be stored and any particular requirements it may have, such as handling or maintenance.

Technical information from the manufacturer can come from a variety of different sources:

* printed or downloadable data sheets

* printed or downloadable user instructions

* manufacturers' catalogues or brochures

* manufacturers' websites.

Organisational documentation

There is a huge potential list of organisational documentation and paperwork. Examples are outlined in Table 2.2. Visual examples can be seen in Fig 2.13 to 2.17.

Document	Purpose
Timesheet	Record of hours that you have worked and the jobs that you have carried out. This is used to help work out your wages and the total cost of the job.
Day worksheet	These detail work that has been carried out without providing an estimate beforehand. They usually include repairs or extra work and alterations.
Variation order	These are provided by the architect and given to the builder, showing any alterations, additions or omissions to the original job.
Confirmation notice	Provided by the architect to confirm any verbal instructions.
Daily report or site diary	These record things that might affect the project like detailed weather conditions, late deliveries or site visitors.
Orders and requisitions	These are order forms, requesting the delivery of materials.
Delivery notes	These are provided by the supplier of materials as a list of all materials being delivered. These need to be checked against materials actually delivered. The buyer will sign the delivery note when they are happy with the delivery.
Delivery record	These are lists of all materials that have been delivered on site.
Memorandum	These are used for internal communications and are usually brief.
Letters	These are used for external communications, usually to customers or suppliers.
Fax	Even though email is commonly used, the industry is still in favour of using faxes, as they provide an exact copy of an original document.

Table 2.2

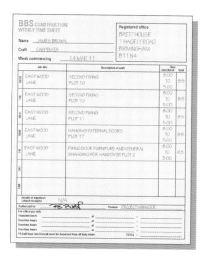

Figure 2.13 Timesheet

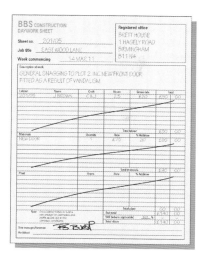

Figure 2.14 Day worksheet

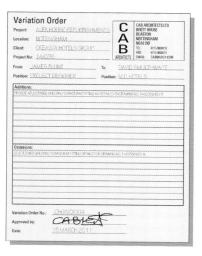

Figure 2.15 Variation order

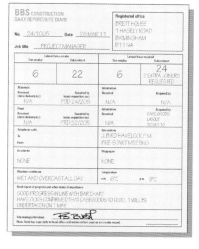

Figure 2.16 Confirmation notice

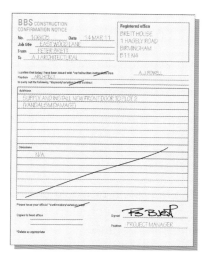

Figure 2.17 Daily report or site diary

Training and development records

Training and development is an important part of any job, as it ensures that employees have all the skills and knowledge that they need to do their work. Most medium to large employers will have training policies that set out how they intend to do this.

To make sure that they are on track and to keep records they will have a range of different documents. These will record all the training that an employee has undertaken.

Training can take place in a number of different ways:

* induction

* toolbox talks

* in-house training

* specialist training

* training or education leading to formal qualifications.

Scales used to produce construction drawings

When the plans for individual buildings or construction sites are drawn up they have to be scaled down so that they will fit on a manageable size of paper. It is important to remember that drawings are not sketches and that they are drawn to scale. This means that they are:

* exact and accurate

* in proportion to the real construction.

You can work out the dimensions by using the scale rule when measuring the drawings. There are several common scales used and the measurement is usually metric:

* 1:2500 – the drawing is 2,500 times smaller than the real object

* 1:100 – the drawing is 100 times smaller than the real object

* 1:50 – the drawing is 50 times smaller than the real object

* 1:20 – the drawing is 20 times smaller than the real object

* 1:10 – the drawing is 10 times smaller than the real object

* 1:5 – the drawing is 5 times smaller than the real object

* 1:2 – the drawing is 2 times smaller than the real object (also called 'half full size').

These drawings would clearly show the dimensions and these would be the actual measurements required (not the scaled-down measurements).

Selecting information

Various documents will provide you with the information you will need. The following examples show you how this works.

Location drawings and specifications

Location drawings are also known as block plans or site plans. The block plan shows the site presented as if you are looking down from above. It will show you where the site is in relation to other buildings and landmarks, such as roads.

The site plan will give you better detail of the site itself. It will contain measurements of the exact dimensions of the plot of land. It will also show you the routes of services and drainage.

The specifications are produced alongside the location drawings of the site. After giving you a brief description of the site, which includes the address, you will find:

* information about any services running into and through the site and whether or not they are connected or need to be connected

* whether there are restrictions about access or working hours on site

* materials that will be needed in order to carry out jobs on site (this will contain quite a lot of detail and will tell you what types of materials are needed, their size, quality and other technical details)

* information about the required workmanship, including what work needs to be done, what quality of work is expected and what the final finish should look like.

Schedules

Schedules can be quite big documents if you are working on a large site with lots of tasks to be done. There is usually a schedule for each different type of job. The schedules are there to record design information. They ensure that you do not accidentally use the wrong components or fittings.

Schedules cover all types of job, such as the types of doors, windows, joinery, heating components and other specific features.

Schedules usually have suitable drawings and usually a floor plan showing where the different features will appear.

DETERMINE QUANTITIES OF MATERIALS

Working out the quantity and cost of resources that are needed to do a particular job is, perhaps, one of the most difficult tasks. In most cases you or the company you work for will be asked to provide a price for the work. It is generally accepted that there are three ways of doing this:

* **An estimate** is an approximate cost produced from the information available before construction begins.

* **A quotation** is a fixed price.

* **A tender** is a bid for the job at your price.

These three ways of costing are very different and each of them has its own problems.

Checking deliveries of building materials

It is important that all deliveries are thoroughly checked as they arrive. Your suppliers will need to have access to the site. It is important to inform suppliers whether the site has any access problems. Construction materials usually arrive on site on large and heavy lorries, so always check if the ground they will have to cross is soft or uneven and warn them if this is the case. Trees and overhead wires could also be a problem, as could finding space to reverse and turn the lorry.

If you are expecting a delivery to arrive, you should be prepared for it. This means ensuring there is:

* clear access to the site

* somewhere ready to store the materials and equipment being delivered

* enough help on hand to assist moving the materials from the delivery point to the storage area.

There are two documents that you will need in order to check the delivery:

* your order, which is a purchase order or confirmation of the materials and equipment that you have ordered from the supplier

* a delivery note, which should be handed to you by the delivery driver.

The first thing to do is check that the two documents match. If they do then what is on the delivery truck should be what you ordered. You now need to check and tick off each item on the delivery note. It needs to be:

* the right specification
* the right size

* the right quantity
* undamaged.

You should not accept items that do not match these four points. You should not sign the delivery ticket or delivery note unless you are satisfied with what has been delivered.

Methods used to estimate quantities

Numerous factors determine the cost of a construction project, whatever its size, but getting the best price on the ideal quantities should not be guesswork. Some possible considerations, according to the size and significance of the project, would be:

* the availability of labour and materials

* the lead time on materials

* the economy and borrowing rates

* the time of year

* cost of plant

* the duration of the project – really large projects can go on for years.

Obviously experience means that you can more quickly estimate the quantities of materials that will be needed on particular (small) construction projects. This is also true of working out the best place to buy materials and how much the labour costs will be to get the job finished.

Many businesses will use the *Hutchins UK Building Blackbook* (published by Franklin-Andrews), which provides a construction cost guide. It breaks down all types of work and shows an average cost for each of them.

Computerised estimating packages are available, which will give a comprehensive detailed estimate that looks very professional. This will also help to estimate quantities and timescales.

The alternative is of course to carry out a numerical calculation. So it is important to have the right resources upon which to base these calculations. These could be working drawings, schedules or other documents.

Usually this involves making additions, subtractions, multiplications and divisions. In order to work out the amount of materials you will need for a construction project you will need to know some basic information:

* What does the job entail? How complex is it, and how much labour is required?

* What materials will be used?

* What are the costs of the materials?

Measurement

The standard unit for measurement is the metre (m). There are 100 centimetres (cm) and 1,000 millimetres (mm) in a metre. It is important to remember that drawings and plans have different scales, so these need to be converted to work out quantities of materials.

The most basic thing to work out is length, from which you can calculate perimeter, area and then volume, capacity, mass and weight, as can be seen in Table 2.3.

Measurement	Explanation
Length	This is the distance from one end to the other. This could be measured in metres or milimetres, depending on the job.
Perimeter	This is the total distance around the outside of a shape. For example, you might need to know the length of the perimeter around a site to work out how much security fencing you need before work starts. You can work out the perimeter by adding the lengths of each side of the shape together. For most jobs, perimeter will be measured in metres (see Fig 2.20).
Area	This is the amount of surface a shape covers. For example you might need to work out the area of a room or a wall to calculate what quantity of materials you will need. You can work out the area of a room by measuring the length and the width of the room and multiplying the two figures together. You can work out the area of a wall by measuring the length and the height of the wall and multiplying the two figures together. For most jobs, area will be measured in square metres (m^2) (see Fig 2.21).
Volume and capacity	This shows how much space is taken up by an object or room. You can work out the volume of a room by multiplying the width by the length and then by the height. For most jobs, volume will be measured in cubic metres (m^3). Capacity works in exactly the same way as volume, but instead of showing the figure as cubic metres you may show it as litres (l). This is ideal if you are trying to work out the capacity of a water tank or a garden pond.
Mass or weight	Mass is measured usually in kilograms or in grams. Mass is the actual weight of a particular object, such as a brick.

Table 2.3

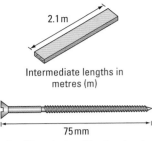

2.1 m

Intermediate lengths in
metres (m)

75 mm

Small lengths in millimetres (mm)

Figure 2.18 Length in metres
and millimetres

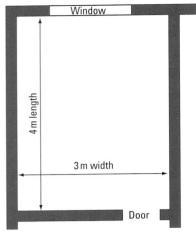

Window

4 m length

3 m width

Door

Figure 2.19 Measuring area and
perimeter

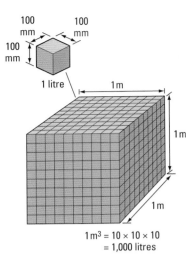

100
mm

100
mm

100
mm

1 litre

1 m

1 m

1 m

$1 m^3 = 10 \times 10 \times 10$
$= 1,000$ litres

Figure 2.20 Relationship between
volume and capacity

Formulae

These can appear to be complicated, but using formulae is essential for working out quantities of materials. Each of the formulae is related to different shapes. In construction you will often have to work out quantities of materials needed for odd shaped areas.

Area

To work out the area of a triangular shape, you use the following formula:

$$\text{Area (A)} = \text{Base (B)} \times \frac{\text{Height (H)}}{2}$$

So if a triangle has a base of 4.5 and a height of 3.5 the calculation is:

$$4.5 \times \frac{3.5}{2}$$

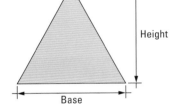

Height

Base

Figure 2.21 Triangle

$$\text{Or } 4.5 \times 3.5 = \frac{15.75}{2} = 7.875 \, m^2$$

Height

If you want to work out the height of a triangle you switch the formulae around. To give:

$$\text{Height} = 2 \times \frac{\text{Area}}{\text{Base}}$$

Perimeter

To work out the perimeter of a rectangle we use the formula:

$$\text{Perimeter} = 2 \times (\text{Length} + \text{Width})$$

It is important to remember this because you need to count the length and the width twice to ensure you have calculated the total distance around the object.

Circles

To work out the circumference or perimeter of a circle you use the formula:

$$\text{Circumference} = \pi \, (\text{pi}) \times \text{diameter}$$

π (pi) is always the same for all circles and is 3.142.

Diameter is the length of the widest part and is twice the radius.

If we know the circumference and need to work out the diameter of the circle the formula is:

$$\text{Diameter} = \frac{\text{circumference}}{\pi \text{ (pi)}}$$

For example if a circle has a circumference of 15.39m then to work out the diameter:

$$\frac{15.39}{3.142} = 4.89\,\text{m}$$

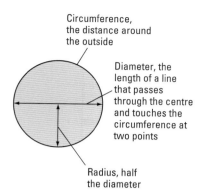

Circumference, the distance around the outside

Diameter, the length of a line that passes through the centre and touches the circumference at two points

Radius, half the diameter

Figure 2.22 Parts of a circle

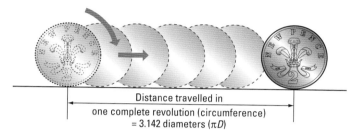

Distance travelled in one complete revolution (circumference) = 3.142 diameters (πD)

Figure 2.23 Relationship between circumference and diameter

Complex areas

Land, for example, is rarely square or rectangular. It is made up of odd shapes. Never be overwhelmed by complex areas, as all you need to do is break them down into regular shapes.

By accurately measuring the perimeter you can then break down the shape into a series of triangles or rectangles. All that is then necessary is to work out the area of each of the shapes within the overall shape and then add them together.

Shape		Area equals	Perimeter equals
Square		AA (or A multiplied by A)	4A (or A multiplied by 4)
Rectangle		LB (or L multiplied by B)	2(L + B) (or L plus B multiplied by 2)
Trapezium		$\dfrac{(A + B)H}{2}$ (or A plus B multiplied by H then divided by 2)	A + B + C + D

Shape		Area equals	Perimeter equals
Triangle	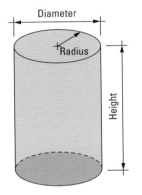	$\dfrac{BH}{2}$ (or B multiplied by H and then divided by 2)	A + B + C
Circle		πR^2 (or Pi (3.142) × R × R)	πD or $2\pi R$ (or Pi (3.142) × D or 2 × 3.142 × R)

Table 2.4

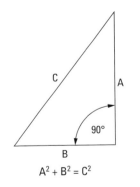

Figure 2.24 Cylinder

Volume

Sometimes it is necessary to work out the volume of an object, such as a cylinder or the amount of concrete needed. All that needs to be done is to work out the base area and then multiply that by the height.

For a concrete area, if a 1.2 m square needs 3 m of height then the calculation is:

$$1.2 \times 1.2 \times 3 = 4.32\,m^3$$

To work out the volume of a cylinder you need to know the base area × the height. The formula is:

$$\pi r^2 \times H$$

So if a cylinder has a radius (r) of 0.8 and a height of 3.5 m then the calculation is:

$$3.142 \times 0.8 \times 0.8 \times 3.5 = 7.038\,m^3$$

Pythagoras

Pythagoras' theorem is used to work out the length of the sides of right angled triangles. It states that:

In all right angled triangles the square of the longest side is equal to the sum of the squares of the other two sides (that is, the length of a side multiplied by itself).

$$A^2 + B^2 = C^2$$

Figure 2.25 Pythagoras' theorem

Measuring materials

Using simple measurements and formulae can help you work out the amount of materials you will need. This is all summarised in the following table.

Material	Measurement
Timber	To work out the linear run of a cubic metre of timber of a given cross sectional area, divide a square metre by the cross sectional area of one piece.
Flooring	To work out the amount of flooring for a particular area metres^2 multiply the width of the floor by the length of the floor.
Stud walling, rafters and joists	Measure the distance that the stud partition will cover then divide that distance by a specified spacing and add 1. This will give you the number of spaces between each stud.
Fascias, barges and soffits	Measure the length and then add 10% for waste; however, this will depend on the nearest standard metric size of timber available.
Skirting, dado, picture rails and coving	You need to work out the perimeter of the room and then subtract any doorways or other openings. Again, add 10% for waste.
Bricks and mortar	Half brick walls use 60 bricks per metre squared and one brick walls use double that amount. You should add 5 per cent to take into account any cutting or damage. For mortar assume that you will need 1 kg for each brick.

Table 2.5

How to cost materials

Once you have found out the quantity of materials necessary you need to find out the price of those materials. It is then simply a case of multiplying those prices by the amount of materials actually needed.

Materials and purchasing systems

Many builders and companies will have preferred suppliers of materials. Many of them will already have negotiated discounts based on their likely spending with that supplier over the course of a year. The supplier will be geared up to supply them at an agreed price.

In other cases builders may shop around to find the best price for the materials that match the specification. It is not always the case that the lowest price is necessarily the best. All materials need to be of a sufficient quality. The other key consideration is whether the materials are immediately available for delivery.

DID YOU KNOW?

Many businesses that fail do so as a result of not working out their costs properly. They may have plenty of work but they are making very little money.

CASE STUDY

LAING O'ROURKE

What you learned at school comes in handy for work

Joshua Richardson is an apprentice in his third year at Laing O'Rourke in Leeds.

'I got an A in Resistant Materials (Design and Technology) at school and then decided it was a good idea to apply for a joinery apprenticeship. I applied for a few apprenticeships and eventually got one with Laing O'Rourke. They gave me a phone interview, and then I had to go to an assessment day where we did things like spatial tests, group activities and a 10-minute presentation. I'm glad I got to do similar tests with my other applications, because if this was the first place I'd done them, maybe I wouldn't have made it.

You had to have GCSEs in Maths, English and Science at A–C. There's some people at college who didn't have these so they have to do Functional Skills on top of everything.

It's important because you do use your Maths and English skills at work every day.

Not only do you need to use your English for college work-based evidence, but also just talking to people. I found that having done talks and presentations at school, it really helped communication skills, and I can talk quite easily with different people now.

With Maths, you definitely need that every day. When you're doing any job, you need to work out how to use a measuring tape, and what kind of Maths you're going to have to use. You have to measure up properly and know each calculation you'll need. It's all on your tape, but you've got to think about it and add it all together, and it's especially important to get it right when you're cutting.

If you're not too good with numbers, you just have to practise it – it's something that can come to you, and not everyone gets it straight away. Every so often, a guy will shout out a couple of measurements and say, "Add that!" So it's not as if you get to pull out your calculator every time.'

It is vital that suppliers are reliable and that they have sufficient materials in stock. Delays in deliveries can cause major setbacks on site. It is not always possible to warn suppliers that materials will be needed, but a well-run site should be able to anticipate needed materials and put the orders in within good time.

Large quantities may be delivered direct from the manufacturer straight to site. This is preferable when dealing with items where consistency is essential.

RELAYING INFORMATION IN THE CONSTRUCTION ENVIRONMENT

Communication is all about passing on accurate information. No matter who you are communicating with, you must understand what is being asked. You also need to be able to give them a clear answer. Sometimes you do not have the information or it is not your decision to make. In these cases you will need to take a message.

Whatever the situation, you always need to be positive and efficient. You also need to be clear. Poor communication and negative communication nearly always lead to confusion, delays and extra costs.

Basic content and requirements for recording a message

Message taking is an important skill. You are the way in which information is passed from one person to another. Providing you remember to pass the message on it may not need to be written down. In many cases, however, messages can be complicated and these do need to be written down. In fact it is a wise plan to record the fact that the message was received in the first place:

* Note the date and time that the message was received.

* Clearly write down the actual message. Someone will have to read this, so make sure it is legible.

* Write down the person's name and their contact details.

If you are taking a message and you are in the site office then there may be a telephone answering pad to hand, or perhaps sticky notes.

If the questions are complicated it will be best to get them to call back or to promise them that the person with the knowledge will call them back.

PRACTICAL TIP

Many people who leave a message will tell you that it is urgent. It is not for you to decide whether it is or not. It is just down to you to pass the message on and mark it urgent if they have said so.

Positive and negative communication

Construction is an industry that relies on communication.

Positive communication means:

* being courteous and respectful when you are talking to others

* being considerate, particularly if the other person is under pressure

* listening to what others are saying

* being clear on key points

* keeping a sense of humour.

Showing these positive communication skills should mean that others will show positive communication towards you.

On the other hand, if you are:

* rude and disrespectful

* unwilling to listen or pay attention

* incapable of making a decision

* bad tempered

then you are communicating in a negative way. This could lead to confusion, arguments and problems.

Clear and effective communication

Communication in all types of work is essential. It needs to be clear and to the point, as well as accurate. Above all it needs to be a two-way process. This means that any communication you have with anyone must be understood. Think before communicating and never assume that someone understands you unless they have confirmed that they do. Negative or poor communication can damage the confidence that others have in you to do your job.

In construction work everything is about time and following strict instructions and specifications. Failing to communicate will always cause confusion, extra cost and delays. In an industry such as this these are unacceptable and very easy to avoid.

Good communication means efficiency and achievement.

COMMUNICATING WITH OTHERS IN THE CONSTRUCTION ENVIRONMENT

Communication can be split into two different types:

* **Verbal communication** includes face-to-face conversations, discussions in meetings or performance reviews and talking on the telephone.

* **Written communication** includes all forms of documents, from letters and emails to drawings and work schedules.

Each of these forms of communication needs to be clear, accurate and designed in such a way as to make sure that whoever has to use it or refer to it understands it.

Communicating in the appropriate way with others

Each construction job will require the services of a team of professionals. They will need to be able to work and communicate with one another. Each has different roles and responsibilities. They can be broken down into three particular groups:

* on site
* off site
* visitors.

These are described in Tables 2.6, 2.7 and 2.8.

On site

Role	Responsibilities
Apprentices	They can work for any of the main building services trades under supervision. They only carry out work that has been specifically assigned to them by a trainer, a skilled operative or a supervisor.
Skilled or trade operative	A specialist in a particular trade, such as bricklaying or carpentry. They will be qualified in that trade, or working towards their qualification
Unskilled operatives	Also known as labourers, these are entry level operatives without any formal training. They may be experienced on sites and will take instructions from the supervisor or site manager.
Building services engineers	They are involved in the design, installation and maintenance of heating, water, electrics, lighting, gas and communications. They work either for the main contractor or the architect and give instruction to building services operatives.
Building services operatives	They include all the main trades involved in installation, maintenance and servicing. They take instruction from the building services engineers and work with other individuals, such as the supervisor and charge-hand.
Sub-contractor	They carry out work on behalf of the main contractor and are usually specialist tradespeople or professionals, such as electricians. Essentially, they provide a service and are contracted to complete their part of the project.
Charge-hand	This person supervises a specific trade, such as carpenters and bricklayers.
Site manager	This person runs the construction site, makes plans to avoid problems and meet deadlines, and ensures all processes are carried out safely. They communicate directly with the client.
Supervisor	The supervisor works directly for the site manager on larger projects and carries out some of the site manager's duties on their behalf.
Health and safety officer	This person is responsible for managing the safety and welfare of the construction site. They will carry out inspections, provide training and correct hazards.

Table 2.6

Off site

Role	Responsibilities
Client	The client, such as a local authority, commissions the job. They define the scope of the work and agree on the timescale and schedule of payments.
Customer	For domestic dwellings, the customer may be the same as the client, but for larger projects a customer may be the end user of the building, such as a tenant renting local authority housing or a business renting an office. These individuals are most affected by any work on site. They should be considered and informed, with a view to them suffering as little disruption as possible.
Architect	They are involved in designing new buildings, extensions and alterations. They work closely with clients and customers to ensure the designs match their needs. They also work closely with other construction professionals, such as surveyors and engineers.

Consultant	Consultants such as civil engineers work with clients to plan, manage, design or supervise construction projects. There are many different types of consultant, all with particular specialisms.
Main contractor	This is the main business or organisation employed to head up the construction work. They organise the on-site building team and pull together all necessary expertise. They manage the whole project, taking full responsibility for its progress and costs.
Clerk of works	This person is employed by the architect on behalf of a client. They oversee the construction work and ensure that it represents the interests of the client and follows agreed specifications and designs.
Quantity surveyor	Quantity surveyors are concerned with building costs. They balance maintaining standards and quality against minimising the costs of any project. They need to make choices in line with Building Regulations. They may work either for the client or for the contractor, and clients and contractors may both have quantity surveyors on site.
Estimator	Estimators calculate detailed cost breakdowns of work based on specifications provided by the architect and main contractor. They work out the quantity and costs of all building materials, plant required and labour costs.
Supplier/wholesaler contracts manager	They work for materials suppliers or stockists, providing materials that match required specifications. They agree prices and delivery dates.

Table 2.7

Visitors

Site visitor	Role and responsibility
Training officers and assessors	These people work for approved training providers. They visit the site to observe and talk to apprentices and their mentors or supervisors. They assess apprentices' competence and help them to put together the paperwork needed to show evidence of their skills.
Building control inspector	This person works for the local authority to ensure that the construction work conforms to regulations, particularly the Building Regulations. They check plans, carry out inspections, issue completion certificates, work with architects and engineers and provide technical knowledge on site.
Water inspector	This person carries out checks of plumbing and drainage systems on construction sites.
Health and Safety Executive (HSE) inspector	An HSE inspector from the local authority can enter any workplace without giving notice. They will look at the workplace, the activities and the management of health and safety to ensure that the site complies with health and safety laws. They can take action if they find there is a risk to health and safety on site.
Electrical services inspector	Inspectors are approved by the National Inspection Council for Electrical Installation Contracting. They check all electrical installation has been carried out in accordance with legislation, particularly Part P of the Building Regulations.

Table 2.8

Maintaining good working relationships

It is important to have a good working relationship with colleagues at work. An important part of this is to communicate in a clear way with them. This helps everyone understand what is going on and what decisions have been made. It also means being clear. Most communication with colleagues will be verbal (spoken). Good communication means:

* cutting out mistakes and stoppages (saving money)

* avoiding delays

* making sure that the job is done right the first time and every time.

Equality and diversity in communication

Equality and diversity is not simply about treating everyone in the same way. It is actually recognising that people are different. Each of us is unique. This could mean that we might have a different culture, be of different ages or follow different religions. It might refer to our marital status or gender, our sexual orientation or our first language.

In all your actions and your communications you should:

* recognise and respect other people's backgrounds

* recognise that everyone has rights and responsibilities

* not harass or be offensive and use language or behaviour that discriminates.

You should also remember that not everyone's first language will be English so they may not understand everything or be able to communicate clearly with you. You might also find that some colleagues may have hearing impairments (or may not hear what you're saying because they are in a noisy environment). It's best to use simple language and check that both you and the person you're communicating with have understood what you need to know.

Figure 2.26 A water inspection

CASE STUDY

LAING O'ROURKE

Using writing and maths in the real world

Gary Kirsop, Head of Property Services says:

'People seem to think that trades are all about your hands, but it's more than that. You're measuring complicated things – all the trades need to have about the same technical level for planning, calculation and writing reports. You need that level to get through your exams for the future too. When you have one day a week in college, but four days a week working with customers in the real world, without communications skills, it would all fall apart. You have to understand that people come from different backgrounds and that they have their own communication modes. Having good GCSEs will really help you get by in the trade.'

TEST YOURSELF

1. If a drawing is at a scale of 1:500, each millimetre in the drawing represents how much on the ground?

   a. 1m

   b. 1mm

   c. 500mm

   d. 500m

2. What is the other term used to describe an orthographic projection?

   a. First angle

   b. Second angle

   c. Third angle

   d. Isometric

3. What is a variation order?

   a. A list of all materials that have been delivered to site

   b. A document showing work that has been carried out without a prior estimate

   c. A document that confirms any verbal instructions

   d. A document provided by the architect to the builder to show any changes to the original job

4. On a drawing, if you were to see the letters FDN, what would that mean?

   a. The signature of the architect

   b. Foundation Design Network

   c. Foundations

   d. Full distance

5. If a drawing is at a scale of 1:5, how many times smaller is the drawing than the real object?

   a. 5 times

   b. 50 times

   c. Half the size

   d. 500 times

6. Which of the following values is pi?

   a. 3.121

   b. 3.424

   c. 3.142

   d. 3.421

7. Which document is used to give detailed sets of requirements that cover the construction, features, materials and finishes?

   a. Work programme

   b. Purchase order

   c. Policy document

   d. Job specification

8. How do you work out the amount of flooring necessary for a room?

   a. Divide width by length

   b. Add width to length

   c. Multiply length by height

   d. Multiply width by length

9. Which individual on a typical site would sign off timesheets?

   a. Architect

   b. Site manager/supervisor

   c. Delivery person

   d. Customer

10. Which are the two main types of communication?

    a. Verbal and written

    b. Telephones and emails

    c. Meetings and memorandum

    d. Plans and faxes

Chapter 3

Unit CSA–L1Core03
KNOWLEDGE OF CONSTRUCTION TECHNOLOGY

LEARNING OUTCOMES

LO1: Know about foundation construction

LO2: Know about floor construction

LO3: Know about wall construction

LO4: Know about roof construction

LO5: Know about utilities and services within construction

LO6: Know about sustainability within construction

INTRODUCTION

The aim of this chapter is to:

* help you understand the range of building materials used within the construction industry

* help you understand their suitability to the construction of modern buildings.

* help you understand the role of sustainability in the construction industry

* help you to be aware of different construction methods.

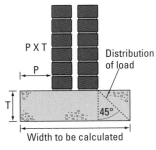

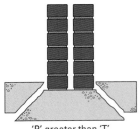

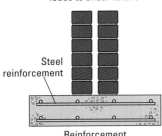

Figure 3.1 Foundation properties

KEY TERMS

Shear failure

– when the load from the superstructure of the building bears down on the foundation. Underneath the foundation the soil will settle and there could be a failure of the soil to support the foundation. This will cause it to crack and part of the building will sink with it.

FOUNDATION CONSTRUCTION

Foundations are a primary element of a building, and form part of the substructure (the element of the building which is below ground and cannot be seen once the structure has been completed). Foundations spread the load of the superstructure (the visible part of the completed building), transferring it to the subsoil ground below. They provide structural stability and help to prevent damage to the building in the event of ground movement. They can range from a concrete strip to pre-cast reinforced concrete-driven piles.

Purpose of foundations

Foundations are designed to counteract factors such as ground movement, which could damage the building. It is important to work out the necessary width of foundation. This depends on the total load of the structure and the load-bearing capacity of the ground or subsoil on which the building is being constructed:

* Wide foundations are used when the construction is on weak ground, or the superstructure will be heavy.

* Narrow foundations are used when the subsoil is capable of carrying a heavy weight, or the building is a relatively light load.

The load that is placed on strip and pad foundations spreads into the ground at 45°. **Shear failure** will take place if the thickness of the foundations is less than the projection of the wall or column face on the edge of the foundations. This is what leads to subsidence (the ground under the structure sinking or collapsing).

As we will see in this section, the depth of the foundation is dependent on the load-bearing capacity of the subsoil. As a general rule of thumb, foundations should be 200 mm to 300 mm thick.

Different types of foundation

Generally there are several different types of foundation, which can be seen in Fig 3.2.

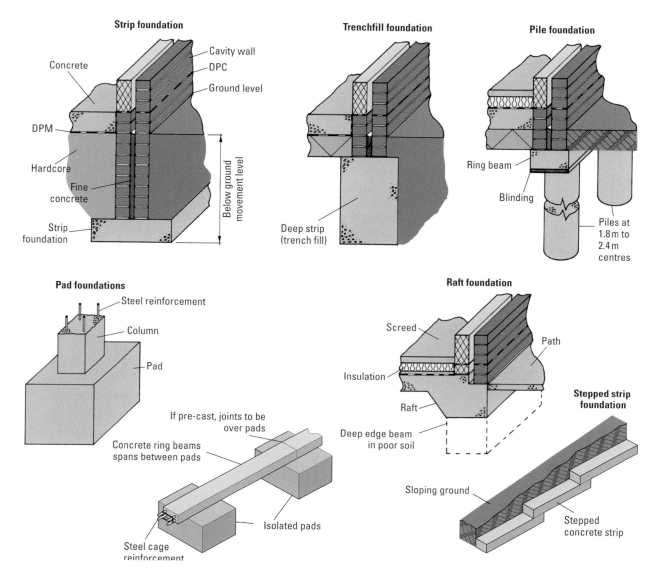

Figure 3.2 Foundation types

* The traditional **strip foundation** is quite narrow and tends to be used for low-rise buildings and dwellings. A thin strip of concrete is laid and then brick or block is built up to the DPC level. These can be reinforced where the ground is weak. They can also be stepped on sloping ground, in order to cut down on the amount of excavation needed. It can also be deep, which uses more concrete but reduces the number of bricks used below ground level. An alternative to deep strip foundations is a trench fill foundation.

* Trench fill foundations are constructed by digging a narrow trench to the foundation depth and then filled with concrete. This reduces the labour and materials required to lay the foundation, as no bricks or blocks have to be laid into the trench. Trench fill:

- reduces the need to have a wide foundation

- reduces construction time

- speeds up the construction of the foundation.

- **Pad foundations** tend to be used for structures that have either a concrete or a steel frame. The pads are placed to support the columns, which transfer the load of the building into the subsoil.

- **Pile foundations** tend to be used for high-rise buildings or where the subsoil is unstable. Holes are bored into the ground and filled with concrete or pre-cast concrete, steel or timber posts are driven into the ground. These piles are then spanned with concrete ring beams with steel reinforcement so that the load of the building is transferred deeper into the ground below. Pile foundations can be short or long depending on how high the building is or how bad the soil conditions are.

- **Raft foundations** are used when there is a danger that the subsoil is unstable. A large concrete slab reinforced with steel bars is used to outline the whole footprint of the building. It has an edge beam to take the load from the walls which is transferred over the whole raft. This means that the building effectively 'floats' on the ground surface on top of the concrete raft.

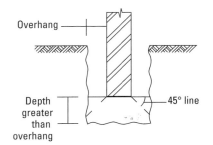

Figure 3.3 Unreinforced strip foundation

Selecting a foundation

One of the first things that a structural engineer will look at when they investigate a site is the nature of the soil, the ground conditions, the likelihood of ground movement and issues such as the water table (where groundwater begins).

Table 3.1 shows different types of subsoil and how they can affect the choice of foundation.

Subsoil type	Characteristics
Rock	High load bearing but there may be cracks or faults in the rock, which could collapse.
Granular	Medium to high load bearing and can be compacted sand or gravel. If there is a danger of flooding the sand can be washed away.
Cohesive	Low to medium load bearing, such as clay and silt. These are relatively stable, but there may be problems with water.
Organic	Low load bearing, such as peat and topsoil. There is also a great deal of air and water present in the soil so organic material must be removed before starting the foundations.

Table 3.1

The ground may move, particularly if the conditions are wet, extremely dry or there are extremes of temperature. Clay, for example, will shrink in the hot summer months and swell up again in the wet winter months. Frost can affect the water in the ground, causing it to expand (frost heave).

Ground movement is also affected by the proximity of trees and large shrubs. They will absorb water from the soil, which can dry out the subsoil. This causes the soil underneath the foundations to collapse. This may not be a problem until the tree or shrub is removed or new ones planted.

As we have seen above, the other key factor when selecting a foundation is the end use of the building:

* Strip foundations – this is the most common and cheapest type of foundation. It is a strip of concrete that runs under the load-bearing walls. The actual depth and width of the strip depends on the ground and the load from the building. They are used for low to medium rise domestic and industrial buildings, as the load bearing requirement will not be high.

* Piled foundations – tend to be used for high rise buildings or where subsoil is unstable if the ground directly under the building is weak or unstable then concrete or steel piles can be driven through this weak ground and into more solid ground beneath it. Reinforced concrete ring beams are placed over the piles as a direct support for the building.

* Raft foundations – these are very expensive and are only ever really used when the ground on which the building is being constructed is very soft. It is also sometimes used when the ground across the area is likely to react in different ways because of the weight of the building. In areas of the UK where there has been mining, for example, raft foundations are quite common, as the building could subside.

Concrete

Concrete is the most common material used for foundations as it is strong and durable. It is usually cast directly on site.

The concrete needs to be poured into the foundation with some care. The size of the foundation will usually determine whether the concrete is actually mixed on site or brought in, in a ready-mixed state, from a supplier. For smaller foundations a concrete mixer and wheelbarrows are usually sufficient. The concrete is then poured into the foundation using a chute (a long trough with a rounded bottom and open ends that directs the concrete to where it is needed).

Concrete consists of both fine and coarse aggregate, along with water.

Aggregates

Aggregates are basically fillers. Coarse aggregate is usually either crushed rock or gravel. The grains are 5mm or larger.

The fine aggregate fills up any gaps between the particles in the coarse aggregate.

Fine aggregate is usually sand that has grains smaller than 5mm.

Cement

Cement is an adhesive or binder. It is Portland stone, crushed, burnt and crushed again and mixed with limestone. The materials are powdered and then mixed together to create a fine powder, which is then fired in a kiln. It may be mixed with other materials for different purposes, such as creating masonry mortar.

Water

Potable water, which is water that is suitable for drinking, should be used when making concrete. The reason for this is that drinkable water has not been contaminated and it does not have organic material in it that could rot and cause the concrete to crack. The water mixes with the cement and then coats the aggregate. This effectively bonds everything together.

Additives

Additives, or admixtures, make it possible to control the setting time and other aspects of fresh concrete allowing you to have greater control over the concrete. They can:

* give you higher strength concrete
* provide protection against degradation of concrete or corrosion of reinforced concrete, which will weaken the structure
* speed up the time the concrete needs to set
* reduce the time the concrete takes to set
* provide protection against cracking as the concrete sets (by preventing shrinkage)
* improve the flow (workability) of the concrete
* improve the finish of the concrete
* provide hot or cold weather protection (a drop or rise in temperature can change the amount of time that concrete needs to set, so these admixtures compensate for that).

You might need this flexibility if, for example, the schedule or weather changes or the job has an unusual specification.

DID YOU KNOW?

Ordinary Portland cement is Portland stone crushed and burnt until all the water disappears. This is taken to site where the water is added again to reconstitute the stone to the form required.

Figure 3.4 Reinforcement using steel bars (or mesh)

Reinforcement

Steel bars or mesh can be used to give the foundation additional strength and support. It can also help to stop the foundation from cracking. Concrete is a good material under direct weight loads, but where concrete foundations are wide and parts of them are under additional tension there is a danger they may crack.

Natural and artificial stone

Both of these products can be placed over the top of plain foundations hardcore fill that is put into the substructure to fill the gap up to the ground floor level hardcore fill that is put into the substructure to fill the gap up to the ground floor level. The synthetic stone weighs far less than natural stone. The other advantage with the synthetic products is that the foundations do not need to be as substantial.

DID YOU KNOW?

Access to some sites, if ready-mixed concrete is being delivered, can sometimes be a problem. A fully loaded mixer truck can weigh over 20 tonnes. Just 6 cubic metres of concrete weighs 14 tonnes.

FLOOR CONSTRUCTION

A floor is a level surface that provides some insulation and carries any loads on it (for example, from furniture) and then to transfer those loads.

Ground floors also have additional purposes. They need to stop moisture from entering the building from the ground. They also need to prevent plant or tree roots from entering the building.

Ground floors

For ground floors there are two options:

* Solid – in contact with the ground

* Suspended – does not touch the ground and spans between walls in the building (effectively there is a void beneath the floor)

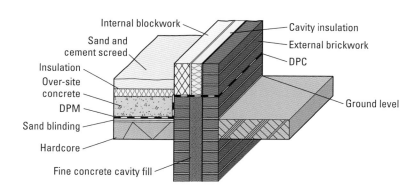

Figure 3.5 Solid ground floors

Floating floor

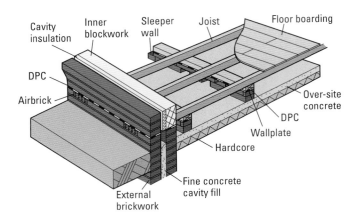

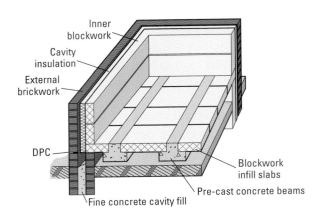

Figure 3.6 Suspended ground floors

The options for ground floors can be complicated because they need to perform several functions. While new builds don't tend to have timber joists and floorboards, extensions to existing buildings usually need to match existing construction styles. Suspended ground floors and traditional timber floors tend to be seen in older buildings. It is far more common to have solid ground floors, or to have timber floors over concrete floors, which are known as floating ground floors.

The key options are outlined in Table 3.2.

Type of floor	Construction and characteristics
Solid	The ground is compacted, and compacted hard core is used as the base, with a binding layer of sand, which is covered with a damp-proof membrane (dpm). A layer of insulation board, usually 100 mm thick, is then placed onto the dpm and concrete is poured on top. To provide a smooth finish for floor finishes a cement and sand screed is applied, usually after the building has been made watertight.
Suspended	Timber – a similar process to a solid ground floor is carried out but then, on top of this, dwarf walls or sleepers are built. These are used to support the timber floor. Air bricks are also added to provide necessary ventilation. Joists are then spaced out along the dwarf walls. A damp-proof course is inserted under the floor joists and then floorboards or sheets placed on top of the joists. Beam and block – concrete beams and lightweight concrete slabs or blocks are used to create the basic flooring. The beams are evenly spaced across the foundation and gaps between the beams are filled with blocks to form the floor. The blocks and beams are then insulated and it is finished off with a cement screed.
Floating	This is a timber construction which goes over the top of a solid concrete floor. Bearers are put down and then the boarding or sheets are fixed to the bearers. The weight of the boards themselves hold them in place.

Table 3.2

Upper floors

Usually for dwellings timber is used for these suspended floors. In industrial buildings beam and block or concrete floor slabs tend to be used.

Timber suspended upper floor

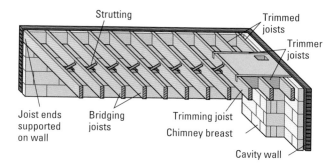

Concrete suspended upper floor

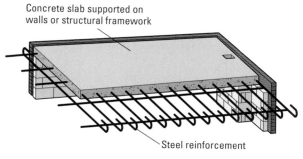

Figure 3.7 Upper floors

For dwellings, bridging joists (horizontal timbers that support the ceiling) are the most common joists used in suspended timber floors. These joists are supported at their ends by load-bearing walls. On the top of the joists, boarding or sheets provide the flooring for the room. Underneath the joists, plasterboard creates the basis of the ceiling for the room below.

When joists have to go into cavity walls (two walls with a hollow space between them) joist hangers are used (U-shaped metal brackets that are used to support the ends of floor joists). There are also complications when joists are in and around stairs and chimney breasts. Bridging joists are used so that these openings are not blocked. Openings in floors require the use of different types of joist called trimmers, trimming and trimmed joists. A trimmed joist is a shortened bridging joist. Any opening in a floor is treated in this way. When the span of a bridging joist exceeds 2 m then struts will be required in line with Building Regulations.

The voids between the floorboards and the plasterboard must be filled with insulation. This not only reduces heat loss, but can also reduce noise.

Concrete suspended floors are usually either cast on site or available as ready-cast units. They are effectively locked into the structure of the building by steel reinforcement. If the concrete floors are being cast on site then **formwork** is needed. Concrete floors are common because they offer greater load bearing capacity, have greater fire resistance and are more sound resistant.

WALL CONSTRUCTION

Walls have a number of different purposes as they:

* hold up the roof

* provide protection against the elements

* keep the occupants of the building warm

* divide the building into rooms, providing privacy and different spaces.

External walls

Many buildings now have cavity walls which means:

* The outside wall is a wet one because it is exposed to the elements outside the building.

* The internal wall is dry but it needs to be kept separate from the outside wall by a cavity.

* The cavity or gap acts as a barrier against damp and also provides some heat insulation.

* The cavity can be completely filled or part-filled depending on the insulation value required by Building Regulations.

Internal walls

Internal walls divide up the space within the building. These do not have all of the demands of the external walls. They are less likely to be load bearing and they do not have to be insulated so are, therefore, thinner. (However, they are commonly insulated in areas such as the toilet or party walls in semi-detached or terrace construction.) They can be brick or block (particularly if they are load bearing), which is then covered with plaster. Alternatively they can be a timber or metal framework, known as stud work, which is covered plasterboard, to form a wall.

Different types of wall construction and structural considerations

In addition to walls being external or internal, they can also be classed as being load bearing or non-load bearing.

Internal walls can be either load bearing or non-load bearing. In both external and internal load bearing walls, any gaps or openings for windows or doors have to be bridged. This is achieved by using either arches or lintels. These support the weight of the wall above the opening.

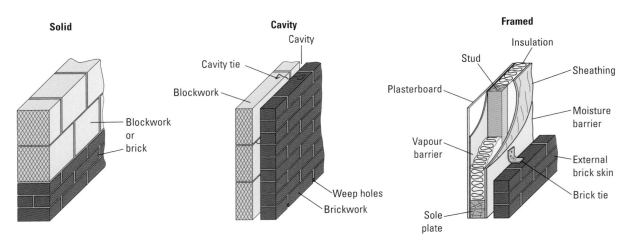

Figure 3.8 Some examples of external wall construction

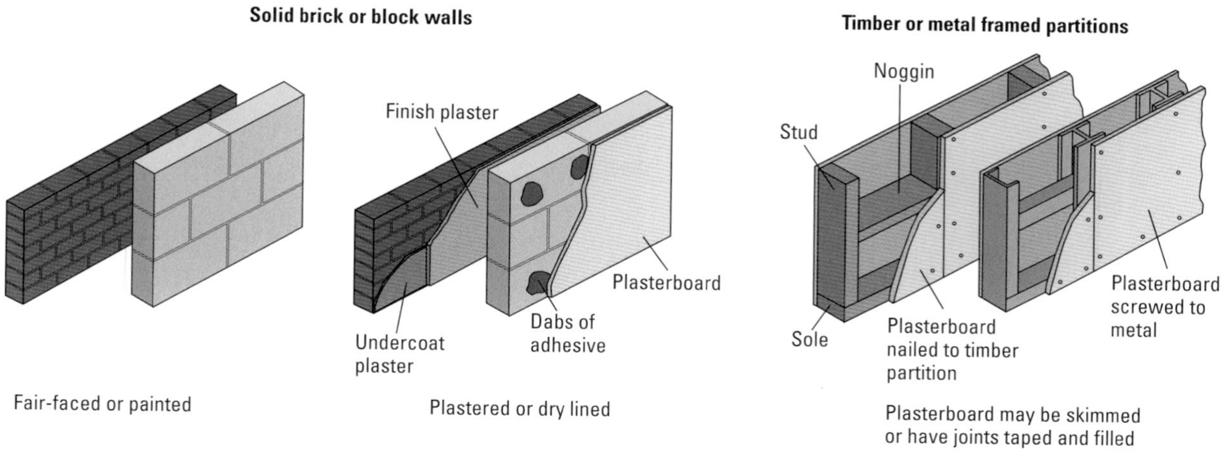

Solid brick or block walls

Timber or metal framed partitions

Fair-faced or painted

Finish plaster

Undercoat plaster

Dabs of adhesive

Plasterboard

Plastered or dry lined

Noggin

Stud

Sole

Plasterboard nailed to timber partition

Plasterboard screwed to metal

Plasterboard may be skimmed or have joints taped and filled

Figure 3.9 Some examples of internal wall construction

Solid walls

In modern builds construction of solid external walls is quite rare. External solid walls tend to be much thinner and made from lightweight blocks in modern builds. They will have some kind of waterproof surface over the top of them, which could be made of render, or plastic, metal or timber cladding.

Cavity walls

As we have seen, cavity walls have an outer and an inner wall and a cavity between them. Usually solid walling, or blockwork, is built up to ground level and then the cavity walling continues to the full height of the building. Alternatively a filled cavity wall is constructed up to ground level. Cavity walls are ideal for most buildings up to medium height.

Many industrial buildings have cavity walls for the lower part of the building and then have insulated steel panels for the top part of the building.

The usual technique is to have brick for the outer wall and an insulating block for the inner wall. The gap or cavity can then be partially filled with an insulation material.

Timber framed walls

Panels made of timber, or in some cases steel, are used to construct walls. They can either be load bearing or non-load bearing and can also be used for the outside of the building or for internal walling. Timber frames are also often clad in brickwork. The panels are solid structures and the spaces between the vertical struts (studs) and the horizontal struts (head or sole plates) are filled with insulation material.

Internal walls

Internal walls are either solid or framed. Solid walls can be made up from blocks or bricks. In many industrial buildings the blocks are actually exposed and can be left in their natural state or painted. In domestic buildings plasterboard is usually bonded to the surface and then plastered over to provide a smoother finish.

It is more common for domestic buildings to have timber or metal-framed internal walls, made from either timber or metal, which are known as stud partitions. These are exactly the same as other framed walling, but will usually have plasterboard fixed to them. They would then receive a skimmed coat of plaster to provide the smooth finish.

Damp-proof membrane (DPM) and damp-proof course (DPC)

Damp-proof membranes are installed under the concrete in ground floors in order to ensure that ground moisture does not enter the building. Effectively it waterproofs the building.

Damp-proof courses are a continuation of the damp-proof membrane. They are built into a horizontal course of either block or brickwork, which is a minimum of 150mm above the exterior ground level. DPCs are also designed to stop moisture from coming up from the ground, entering the wall and then getting into the building. The most common DPC is a polythene sheet damp-proof membrane, which comes in rolls the width of the blockwork or brickwork. In older buildings lead, bitumen or slate would have been used as a DPC.

ROOF CONSTRUCTION

In a country such as the UK, with a great deal of rain and snow, it makes sense for roofs to be pitched. Pitched means built at an angle. The idea is that the rain and snow falls down the angle and off the edge of the roof or into gutters rather than lying on the roof.

This is not to say that all roofs are pitched. In fact many domestic dwelling extensions have flat roofs. A great number of industrial buildings have entirely flat roofs. The problem with a flat roof is that it needs to be able to support itself, but as importantly it needs to be able to carry the additional weight of snow or rain. This means that large flat roofs may have to have steel sections (known as trusses) or even reinforced concrete and beams to increase their load-bearing capacity.

Roofs also provide stability to the walls by tying them together. As we will see, there are several different types of roof. These are usually identified by their pitch or shape.

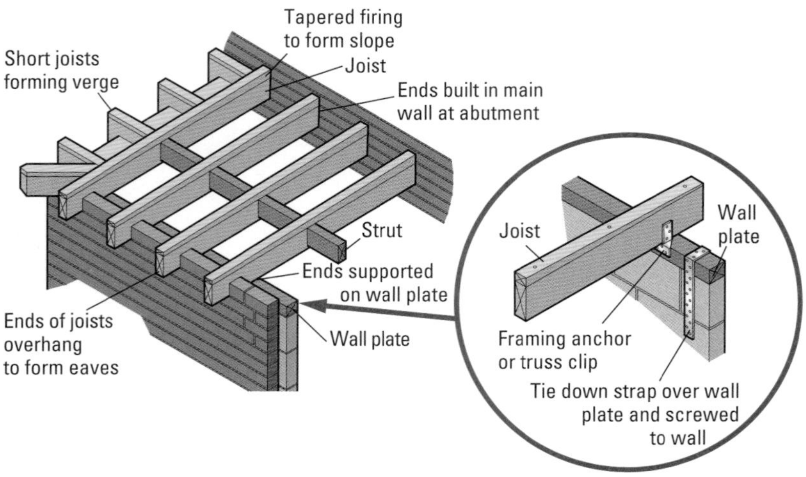

Short joists forming verge

Tapered firing to form slope

Joist

Ends built in main wall at abutment

Strut

Joist

Wall plate

Framing anchor or truss clip

Tie down strap over wall plate and screwed to wall

Ends supported on wall plate

Wall plate

Ends of joists overhang to form eaves

Figure 3.10 Flat roof structure

Types of roof construction

The roof is made up of the rafters and beams, Everything above the framework is regarded as a roof covering, such as slates, tiles and felt. Generally speaking, roofs are either pitched or flat. This will depend on the angle or slope of the roof.

Table 3.3 outlines some of the key characteristics of different types of roof.

Flat	This is a roof that has a slope of less than 10°. Generally flat roofs are used for smaller extensions to dwellings and on garages. Traditionally they would have had bitumen felt, although it is becoming more common for fibreglass to be used.	 Figure 3.11
Mono-pitch	This is a roof that has a single sloping surface but is not fixed to another building or wall. The front and back walls could be different heights, or the other exposed surface of the roof is perpendicular.	 Figure 3.12
Double pitched	This is a roof that has two differently angled slopes. Usually the upper part of the roof has a fairly shallow pitch or slope and the lower part of the roof has a steeper slope.	 Figure 3.13
Couple roof	This is often called gable end and is one of the most common types of roof for dwellings. A gable is a wall with a triangular upper part. This supports the roof in construction using purlins. This means that the roof has two sloping surfaces, which come down from the ridge to the eaves.	 Figure 3.14
Hipped roof	Hipped roofs have slopes on three or four four sides. There are also hipped roofs with single, straight gables.	 Figure 3.15
Lean-to	A lean to is similar to a mono-pitched roof except it is abutted to a wall. The slope is greater than 10°. The higher part of the roof is fixed to a higher wall.	 Figure 3.16

Table 3.3 Different types of roof

Roofing components

Each part of a roof has a specific name and purpose. Table 3.4 explains each of these individual features.

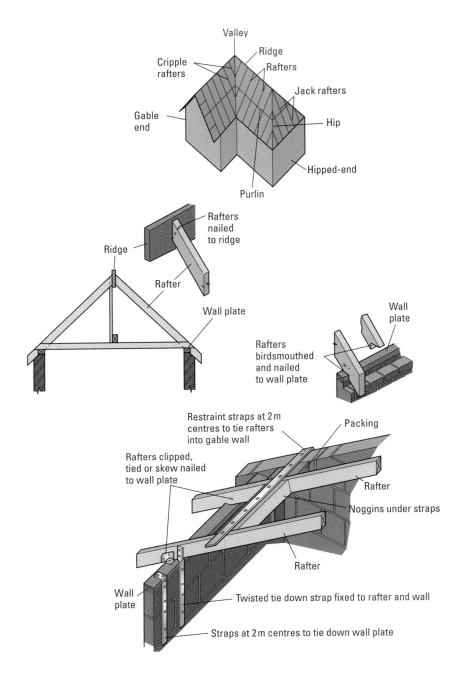

Figure 3.17 Traditional cut roof details

Roof feature	Description
Ridge	This is the top of the roof and the junction of the sloping sides. It is the apex, where the rafters meet.
Purlin	This is a beam that supports the mid-span section of rafters.
Firings	These are angled pieces of timber that are placed on the rafters to create a slope.
Batten	Roof battens are thin strips, usually of wood, which provide a fixing point for either roofing sheets or roof tiles.
Tile	These are artificial products that can be made from clay, concrete or plastic. They are placed in regular, overlapping rows and fixed to the battens.
Fascia	This is a horizontal board that closes and protects the rafter ends and provides a fixing for guttering. It is also decorative as it covers the rafter ends. It is fixed to the ends of the rafters at eaves level and is both a decorative feature and a fixing for rainwater goods.
Wall plate	This is a piece of horizontal timber that is placed at the top of a wall at eaves level. It provides a fixing for joists or rafters.
Bracings	Roof rafters may need to be braced to make them more rigid and stable. These bracings prevent the roof from buckling.
Felt	Roofing felt has two elements – it has a waterproofing agent (bitumen) and what is known as a carrier. The carrier can be either a polyester sheet or a glass fibre sheet. Roofing felt tends to be used for flat roofs and for roofs with a shallow pitch.
Slate	Slate roofing tiles are natural products and are usually fixed to timber battens with double nails. They have a lifespan of between 80 and 100 years.
Flashings	Wherever there is a joint or angle on a roof, a thin sheet of either lead or another waterproof material is added. In the past this tended always to be made from lead. Many different types of flashing can now be used but all have the role of preventing water penetrating into joints. Flashings are normally found where roofs abut a wall or where chimneys protrude through a roof.
Rafter	Roof rafters are the main structural components of the roof. They are the framework. They rest on supporting walls. The rafters are set at an angle on sloped roofs or horizontal on a flat roof.
Apex	The apex is the highest point of the roof, usually the ridge line.
Soffit	Soffits are the lower part, or overhanging part, of the eaves. In other words they are the underside of the eaves.
Bargeboard	This is a functional and ornamental feature, which is fixed to the gable end of a roof in order to hide the ends of roof timbers and to support the verge details.
Eaves	These are the area found at the foot of the rafter. They are not always visible as they can be flush. In modern construction, the eaves have two parts: the visible eaves projection and the hidden eaves projection.

Table 3.4

Roof coverings

There are many different types of materials that can be used to cover the roof. Even tiles and slates come in a wide variety of shapes and sizes, along with colours and different finishes.

In many cases the type of roof covering is determined by the traditional and local styles in the area. Local authorities usually want roof coverings that are not too far from the common style in the area. This does not stop manufacturers from coming up with new ideas, however, which can add benefits during construction and during the use of the building (such as better insulation properties).

Table 3.5 outlines some of the more common types of roof covering and describes their main characteristics and use.

Roof covering	Description
Felt  Figure 3.18	Felt is used as a waterproof barrier. Internal felt is rolled over the top of the rafters. The strips are overlapped to provide a permanent waterproof barrier. They are then battened down and another roof covering, such as slate or tile, placed over the top of them. For flat roofs, felt is used as the external roof covering and is covered in a waterproof material, such as bitumen.
Slate  Figure 3.19	Slate is a flat, natural substance, which is laid onto the battens with each slate tile overlapping the top of the slate in the row directly below it. The slate tiles are either nailed or hooked into place.
Tile 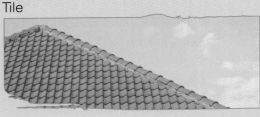 Figure 3.20	There is a huge variety of roofing tiles, made from clay, ceramics or concrete. They are designed and moulded so that they overlap with one another and are fixed to the roof in a similar way to slate tiles.
Metals 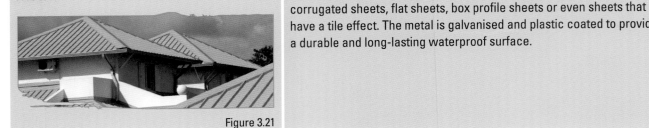 Figure 3.21	There are many different types of metal roof covering, such as corrugated sheets, flat sheets, box profile sheets or even sheets that have a tile effect. The metal is galvanised and plastic coated to provide a durable and long-lasting waterproof surface.

Table 3.5

CASE STUDY

How to impress in interviews

Andrea Dickson and Gillian Jenkins sit on the interview panels for apprenticeship applications at South Tyneside Homes.

'Interviews are all about the three Ps: Preparation, Presentation and Personality.

An applicant should turn up with some knowledge about the apprenticeship programme and the company itself. For example, knowing how long it is, that they have to go to college and to work – don't say, "I was hoping you'd tell me about it"! If they've done a bit of research, it will show through and work in their favour – especially if they can explain why it is that they want to work here.

It sets them up for the interview if they come in smartly dressed. We're not marking on that, but it does show respect for the situation. It's still a formal process and, although we try to make them feel at ease as much as we possibly can, there's no getting away from the fact that they're applying for a job and it is a formal setting.

The interviews are a chance to tell the company about themselves: what they do in their spare time, what their greatest achievements have been and why. Applicants should talk about what interests them; for example, are they really interested in becoming a joiner or is that something their parents want them to do? An apprenticeship has to be something they want to do – if they have enthusiasm for the programme, then they'll fly through it. If not, it's a very long three to four years. Without that passion for it, the whole process will be a struggle; they'll come in late to work and even fail exams.

We also talk to them about any customer service experiences they've had, working in a team, project working (for example, a time you had to complete a task and what steps you took), as well as asking some questions about health and safety awareness.'

SUPPLY OF UTILITIES AND SERVICES

Most but not all dwellings and other structures are connected in some way to a wide range of utilities and services. In the majority of cities, towns and villages structures are connected to key utilities and services, such as a sewer system, potable (drinking) water, gas and electricity. This is not always the case for more remote structures, however.

Whenever construction work is carried out, whether it is on an existing structure or a new build, the supply of utilities and services or the linking up of these parts of the **infrastructure** are very important. Often they will require the services of specialist engineers from the **service provider.**

KEY TERMS

Infrastructure

– these are basic facilities, such as a power supply, a road network and a communication link.

Service provider

– these are companies or organisations that provide utilities, such as gas, water, communications or electricity.

Table 3.6 outlines the main utilities and services that are provided to most structures.

Utility or service	Description
Drainage	Drainage is delivered by a range of water and sewerage companies in the UK. They are responsible for ensuring that surface water can drain away into their system.
Waste water and sewerage	Any waste water and sewage generated by the occupants of a structure needs to have the necessary pipework to link it to the main sewerage system. It is then sent to a sewage treatment works via the pipework. If there is no connection to mains sewerage, the building may have a septic tank, which is a small-scale, self-contained sewage treatment system.
Water	Each structure should be linked to the water supply that provides wholesome, potable drinking water. The pipework linking the structure to the water supply needs to be protected to ensure that backflow from any other source does not contaminate the system.
Gas	Each area has a range of different gas suppliers. This is delivered via a service pipe from the main system into the structure. Areas that do not have access to the main gas supply system use gas contained in cylinders.
Electricity	The National Grid provides electricity to a variety of different electricity suppliers. It is the National Grid that operates and maintains the cabling. There are around 28 million individual electricity customers in the UK.
Communications (telephone, data, cable)	There are several ways in which telecommunications can be linked to a structure. Traditional telephone poles hold up copper cables and not only provide telephone but also internet access to structures. In cities and many of the larger towns poles are being replaced by cables that are fibre optic and run underground. These are then linked to each individual structure.
Ducting (heating and ventilation)	Heating and ventilation engineers install and maintain duct work. The complex systems are known as HVAC. These systems can transfer air for heating or cooling of the structure. The overall system can also provide hot and cold water systems, along with ventilation.

Table 3.6

KEY TERMS

HVAC

– this is an abbreviation for 'heating, ventilation and air-conditioning'. This has been a service provided to many industrial buildings for a number of years, but it is now becoming more common in domestic dwellings, particularly new developments.

SUSTAINABILITY AND INCORPORATING SUSTAINABILITY INTO CONSTRUCTION PROJECTS

Carbon is present in all fossil fuels, such as coal or natural gas. Burning fossil fuels releases carbon dioxide, which is a greenhouse gas linked to climate change.

Energy conservation aims to reduce the amount of carbon dioxide in the atmosphere. The idea is to do this by making buildings better insulated and, at the same time, make heating appliances more efficient. It also means attempting to generate energy using renewable and/or low or zero carbon methods.

According to the government's Environment Agency, sustainable construction is all about using resources in the most efficient way. It also means cutting down on waste on site and reducing the amount of materials that have to be disposed of and put into landfill.

In order to achieve sustainable construction the Environment Agency recommends:

* reducing construction, demolition and excavation waste that needs to go to landfill

* cutting back on carbon emissions from construction transport and machinery

* responsibly sourcing materials

* cutting back on the amount of water that is wasted

* making sure construction does not have an impact on biodiversity.

Sustainable construction and incorporating it into construction projects

Recently the idea of sustainable construction has focused on ensuring that the building is not only of good quality and affordable, but also that it is efficient.

Sustainable construction also means having the least negative environmental impact. So this means minimising the use of raw materials, energy, land and water. This is not only during the construction phase but also for the lifetime of the building.

Finite and renewable resources

We all know that resources such as coal and oil will eventually run out. These are examples of finite resources.

Oil is not just used as fuel – it is in plastic, dyes, lubricants and textiles. All of these are used in the construction process.

Renewable resources are those that are produced either by moving water, the sun or the wind. They include materials that come from plants, such as biodiesel, or the oils used to make adhesives.

The construction process itself is only part of the problem. It is also the longer term impact and demands that the building will have on the environment. This is why there has been a drive towards sustainable homes and there is a Code for Sustainable Homes (an environmental assessment method for rating and certifying the performance of new homes).

Figure 3.22 Most modern new-builds follow sustainable principles

Construction and the environment

In 2010 construction, demolition and excavation produced 20 million tonnes of waste that had to go into landfill. The construction industry is also responsible for most illegal fly tipping (illegally dumping waste). In any year the Environment Agency responds to around 350 serious pollution incidents caused as a result of construction.

Regardless of the size of the construction job, everyone working on the project is responsible for the impact they have on the environment. Good site layout, planning and management can help reduce these problems.

Sustainable construction helps to encourage this because it means managing resources in a more efficient way, reducing waste and reducing your **carbon footprint.**

Architecture and design

The Code for Sustainable Homes Rating Scheme was introduced in 2007. Many local authorities have instructed their planning departments to encourage sustainable development. This begins with the work of the architect who designs the building.

Local authorities ask that architects and building designers:

* ensure the land is safe for development – that if it is contaminated this is dealt with first

* ensure there is access to and protection of the natural environment – this helps ensure biodiversity and tries to create open spaces for local people

* reduce the negative impact on the local environment – any buildings keep noise, air, light and water pollution down to a minimum

* conserve natural resources and cut back carbon emissions – this includes use of energy, materials and water during construction and the life of the building

* ensure comfort and security – good access, close to public transport, safe parking and protection against flooding.

Figure 3.23 Sustainable developments aim to be pleasant places to live

Using locally managed resources

The construction industry imports nearly 6 million cubic metres of sawn wood each year. However there is plenty of scope to use the many millions of cubic metres of timber produced in managed forests in the UK, particularly in Scotland.

Local timber can be used for a wide variety of different construction projects:

* softwood – including pines, firs, larch and spruce – for panels, decking, fencing and internal flooring

* hardwood – including oak, chestnut, ash, beech and sycamore – for a wide variety of internal joinery.

Using local materials reduces transportation costs and time, minimises the project's carbon footprint and means that there is less chance for the materials to be damaged in transit.

Eco-friendly, sustainable manufactured products and environmentally resourced timber

There are now many suppliers that offer sustainable building materials as a green alternative. Some tiles, for example, are now made from recycled plastic bottles and stone particles.

Figure 3.24 Window frames made from timber

There is now a National Green Specification database of all environmentally friendly building materials. This provides a checklist where it is possible to compare specifications of sustainable products to traditionally manufactured products, such as bricks.

Simple changes can be made, such as using timber or ethylene-based plastics instead of UPVC window frames to ensure a building uses more sustainable materials.

As we have seen, finding locally managed resources, such as timber, makes sense in terms of cost and in terms of protecting the environment. There are always alternatives to the use of traditional resources that could affect the environment.

The Timber Trade Federation produces a timber certification system. This ensures that wood products are labelled to show that they are produced in sustainable forests.

Around 80 per cent of all the softwood used in construction comes from Scandinavia or Russia. Another 15 per cent comes from the rest of Europe, or even North America. The remaining 5 per cent comes from tropical countries, and is usually sourced from sustainable forests.

Alternative methods of building

The most common type of construction is, of course, brick and blockwork. However there are plenty of other options:

* timber frame – using green oak

* insulated concrete formwork – where a polystyrene mould is filled with reinforced concrete

DID YOU KNOW?

www.recycledproducts.org.uk has a long list of recycled surfacing products, such as tiles, recycled wood and paving and details of local suppliers.

Figure 3.25 Timber Certification System

* structural insulated panels – where buildings are made up of rigid building boards rather like huge sandwiches

* modular construction – this uses similar materials and techniques to standard construction, but the units are built off site and transported ready-constructed to building site where they are connected together.

Figure 3.26 Green roofing

Figure 3.27 Flooring made from cork

KEY TERMS

Biodegradable

– this material will more easily break down when it is no longer needed. This breaking down process is done by micro-organisms.

Organic

– these are natural substances, usually extracted from plants.

There are alternatives to traditional flooring and roofing, all of which are greener and more sustainable. Green roofing has become an increasing trend in recent years. Metal roofs made of steel, aluminium and copper use a high percentage of recycled material. Solar roof shingles, or solar roof laminates, while expensive, decrease the cost of electricity and related heating costs of the dwelling. Some buildings even have a waterproof membrane, which is covered with a growing medium and planted with vegetation like sedum plants. This provides additional insulation, absorbs air pollution, helps to collect and process rainwater and keeps the roof surface temperature down.

Just as roofs are becoming greener, so too are the options for flooring. The use of bamboo, eucalyptus and cork is becoming more common. A new version of linoleum has been developed with **biodegradable**, **organic** ingredients. Some buildings are also using sustainable alternatives to traditional timber floorboards and joists, and these can be coloured, stained or patterned.

An increasing trend has been for what is known as off-site manufacture (OSM). European OSM businesses, particularly those in Germany, have built over 100,000 houses. The entire house is manufactured in a factory and then assembled on site. Walls, floors, roofs, windows and doors with built-in electrics and plumbing, all arrive on a lorry. Some manufacturers even offer completely finished dwellings, including carpets and curtains. Many of these modular buildings are designed to be far more energy efficient than traditional brick and block constructions. Many come ready fitted with heat pumps, solar panels and triple-glazed windows.

Energy efficiency and incorporating it into construction projects

Energy efficiency is all about using less energy to provide the same level of output. Governments are working towards the world's energy needs by 30 per cent before 2050. This means producing more energy efficient buildings. It also means using energy efficient methods to produce materials and resources needed to construct buildings.

Building Regulations

In terms of energy conservation, the most important UK law is the Building Regulations 2010, particularly Part L. The Building Regulations:

* list the minimum efficiency requirements

* provide guidance on compliance, the main testing methods, installation and control

* cover both new dwellings and existing dwellings.

A key part of the regulations is the Standard Assessment Procedure (SAP), which measures or estimates the energy efficiency performance of buildings.

Local planning authorities also now require that all new developments generate at least 10 per cent of their energy from renewable sources. This means that each new project has to be assessed one at a time.

Energy conservation

By law, each local authority is required to reduce carbon dioxide emissions and to encourage the conservation of energy. This means that everyone has a responsibility in some way to conserve energy:

* Clients, along with building designers, are required to include energy efficient technology in the build.

* Contractors and sub-contractors have to follow these design guidelines. They also need to play a role in conserving energy and resources when working on site.

* Suppliers of products are required by law to provide information on energy in the production of their products.

In addition, new energy efficiency schemes and Building Regulations cover the energy performance of buildings. Each new build is required to have an Energy Performance Certificate. This rates a building's energy efficiency from A (which is very efficient) to G (which is very inefficient).

Some building designers have also begun to adopt other voluntary ways of attempting to protect the environment. These include BREEAM, which is an environmental assessment method, and the Code for Sustainable Homes, which is a certification of sustainability for new builds.

energy saving trust

Figure 3.28 The Energy Saving Trust encourages builders to use less wasteful building techniques and more energy efficient construction

High, low and zero carbon

When we look at energy sources, we consider their environmental impact in terms of how much carbon dioxide they release. Accordingly, energy sources can be split into three different groups:

* high carbon – those that release a lot of carbon dioxide

* low carbon – those that release some carbon dioxide

* zero carbon – those that do not release any carbon dioxide.

Some examples of high carbon, low carbon and zero carbon energy sources are given in Table 3.7.

High carbon energy source	Description
Natural gas or LPG	Piped natural gas or liquid petroleum gas stored in bottles
Fuel oils	Domestic fuel oil, such as diesel
Solid fuels	Coal, coke and peat
Electricity	Generated from non-renewable sources, such as coal-fired power stations
Low carbon energy source	
Solar thermal	Panels used to capture energy from the sun to heat water
Solid fuel	Biomass such as logs, wood chips and pellets
Hydrogen fuel cells	Convert chemical energy into electrical energy
Heat pumps	Convert low temperature heat into higher temperature heat
Combined heat and power (CHP)	Generates electricity as well as heat for water and space heating
Combined cooling, heat and power (CCHP)	A variation on CHP that also provides a basic air conditioning system
Zero carbon energy	
Electricity/wind	Uses natural wind resources to generate electrical energy
Electricity/tidal	Uses wave power to generate electrical energy
Hydroelectric	Uses the natural flow of rivers and streams to generate electrical energy
Solar photovoltaic	Uses solar cells to convert light energy from the sun into electricity

Table 3.7

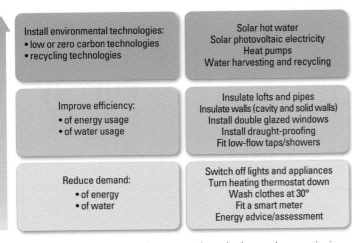

Figure 3.29 Working towards reducing carbon emissions

It is important to try to conserve non-renewable energy so that there will be sufficient fuel for the future. The idea is that finite sources of fuel should last as long as is necessary to completely replace it with renewable sources, such as wind or solar energy.

Alternative heating sources

There are several new ways in which we can harness the power of water, the sun and the wind to provide us with new heating sources. All of these systems are considered to be far more energy efficient than traditional heating systems, which rely on gas, oil, electricity or other fossil fuels.

Solar thermal

At the heart of this system is the solar collector, which is often referred to as a solar panel. The idea is that the collector absorbs the sun's energy, which is then converted into heat. This heat is then applied to the system's heat transfer fluid.

The system uses a differential temperature controller (DTC) that controls the system's circulating pump when solar energy is available and there is a demand for water to be heated.

In the UK, due to the lack of guaranteed solar energy, solar thermal hot water systems often have an auxiliary heat source, such as an immersion heater.

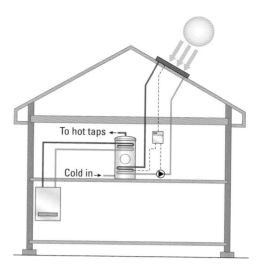

Figure 3.30 Solar thermal hot water system

Biomass (solid fuel)

Biomass stoves burn either pellets or logs. Some have integrated hoppers that transfer pellets to the burner. Biomass boilers are available for pellets, woodchips or logs. Most of them have automated systems to clean the heat exchanger surfaces. They can provide heat for domestic hot water and space heating.

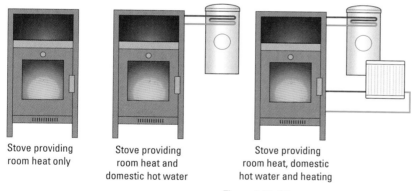

Stove providing room heat only

Stove providing room heat and domestic hot water

Stove providing room heat, domestic hot water and heating

Figure 3.31 Biomass stoves output options

Heat pumps

Heat pumps convert low temperature heat from air, ground or water sources to higher temperature heat. They can be used in ducted air or piped water **heat sink** systems.

There are a variety of different arrangements for each of the three main systems:

* Air source pumps operate at temperatures down to minus 20°C. They have units that receive incoming air through an inlet duct.

* Ground source pumps operate on **geothermal** ground heat. They use a sealed circuit collector loop, which is buried either vertically or horizontally underground.

KEY TERMS

Heat sink

– this is a heat exchanger that transfers heat from one source into a fluid, such as in refrigeration, air-conditioning or the radiator in a car.

Geothermal

– relating to the internal heat energy of the earth.

* Water source systems can be used where there is a suitable water source, such as a pond or lake.

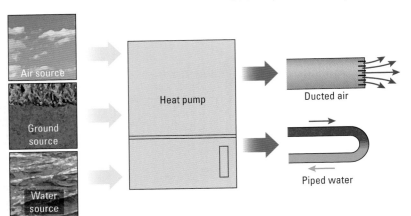

Figure 3.32 Heat pump input and output options

The heat pump system's efficiency relies on the temperature difference between the heat source and the heat sink. Special tank hot water cylinders are part of the system, giving a large surface-to-surface contact between the heating circuit water and the stored domestic hot water.

Combined heat and power (CHP) and combined cooling heat and power (CCHP) units

These are similar to heating system boilers, but they generate electricity as well as heat for hot water or space heating (or cooling). The heart of the system is an engine or gas turbine. The gas burner provides heat to the engine when there is a demand for heat. Electricity is generated along with sufficient energy to heat water and to provide space heating.

CCHP systems also incorporate the facility to cool spaces when necessary.

Wind turbines

Freestanding or building-mounted wind turbines capture the energy from wind to generate electrical energy. The wind passes across rotor blades of a turbine, which causes the hub to turn. The hub is connected by a shaft to a gearbox. This increases the speed of rotation. A high speed shaft is then connected to a generator that produces the electricity.

Solar photovoltaic systems

A solar photovoltaic system uses solar cells to convert light energy from the sun into electricity. The solar cells are usually made of silicon and are semi-conductors. The sunlight hits the solar cells and photons are absorbed. This causes negatively charged electrons in the cell to detach from their atoms and flow through the cell to create electricity. The electricity is direct current (dc). The dc current is then converted by an inverter to alternating current (ac), which is the type of current used for mains electricity.

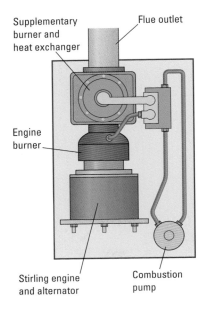

Figure 3.33 Example of a MCHP (micro combined heat and power) unit

Energy ratings

Energy rating tables are used to measure the overall efficiency of a dwelling, with rating A being the most energy efficient and rating G the least energy efficient.

Alongside this is an environmental impact rating (see Fig 3.38). This measures the dwelling's impact on the environment in terms of how much carbon dioxide it produces. Again, rating A is the highest, showing it has the least impact on the environment, and rating G is the lowest.

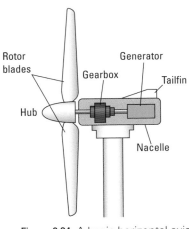

Figure 3.34 A basic horizontal axis wind turbine

A standard assessment procedure (SAP) is used to place the dwelling on the energy rating table. This will take into account:

* the date of construction, the type of construction and the location

* the heating system

* insulation (including cavity wall)

* double glazing.

The ratings are used by local authorities and other groups to assess the energy efficiency of new and old housing, and must be provided to potential purchasers when houses are sold.

Preventing heat loss

Most old buildings are under-insulated and would benefit from additional insulation, whether this is ceilings, walls or floors.

The measurement of heat loss in a building is known as the U Value. It measures how well parts of the building transfer heat. Low U Values represent high levels of insulation. U Values are becoming more important as they form the basis of energy and carbon reduction standards.

By 2016 all new housing is expected to be Net Zero Carbon. This means that the building should not be contributing to climate change.

Many of the guidelines are now part of Building Regulations (Part L). They cover:

* insulation requirements

* openings, such as doors and windows

* solar heating and other heating

* ventilation and air conditioning

* space heating controls

* lighting efficiency

* air tightness.

Building design

UK households spend £2.4bn every year just on lighting. One of the ways of tackling this cost is to use energy saving lights, but also to maximise natural lighting. For the construction industry this means:

* increased window size

* orientating building angles to make the most of sunlight – south facing windows maximise sunlight in winter and limit overheating in the summer

* considering window design by using windows with a variety of different types of opening to allow ventilation.

Solar tubes are another way of increasing light. These are small domes on the roof, which collect sunlight and then direct it through a tube (which is reflective). It is then directed through a diffuser in the ceiling to spread light into the room.

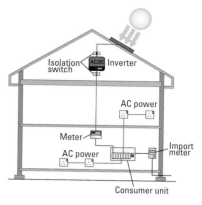

Figure 3.35 A basic solar photovoltaic system

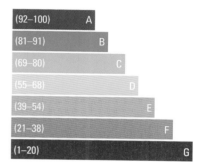

Figure 3.36 SAP energy efficiency rating table. The ranges in brackets show the percentage energy efficiency for each banding

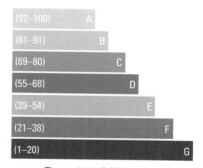

Figure 3.37 SAP environmental impact rating table

TEST YOURSELF

1. In which of the following types of buildings is a traditional strip foundation used?

   a. High rise

   b. Medium rise

   c. Low rise

   d. Industrial buildings

2. Which of the following is a reason for using a raft foundation?

   a. The subsoil is rock

   b. The subsoil is unstable

   c. The subsoil is stable

   d. The access to the site allows it

3. What holds down a floating floor?

   a. Nails and screws

   b. Adhesives

   c. Blocks

   d. Its own weight

4. What is another term for formwork?

   a. Shuttering

   b. Cavity

   c. Joist

   d. Boarding

5. What is the minimum distance the DPC should be above ground level?

   a. 50 mm

   b. 100 mm

   c. 150 mm

   d. 200 mm

6. A roof is said to be flat if it has a slope of less than how many degrees?

   a. 5

   b. 10

   c. 15

   d. 20

7. What shape is the upper part of a gable end?

   a. Rectangular

   b. Semi-circular

   c. Square

   d. Triangular

8. What do you call the horizontal timber that is placed at the top of a wall at eaves level in a roof, to hold the ends of joists or rafters?

   a. Fascia

   b. Bracings

   c. Wall plate

   d. Batten

9. What happens to the majority of construction demolition and excavation waste?

   a. It is buried on site

   b. It is burned

   c. It goes into landfill

   d. It is recycled

10. Which part of the Building Regulations 2010 requires the construction industry to consider and use energy efficiently?

    a. Part B

    b. Part D

    c. Part K

    d. Part L

Chapter 4

Unit CSA–L1Occ16
PREPARE BACKGROUND SURFACES AND PLASTER MATERIALS

LEARNING OUTCOMES

LO1/2: Know how to and be able to prepare for the preparation of background surfaces

LO3/4: Know how to and be able to prepare background surfaces

LO5/6: Know how to and be able to prepare for and mix plaster materials

INTRODUCTION

The aims of this chapter are to:

* show you how to prepare background surfaces for plastering

* help you to choose, prepare and mix the appropriate materials for the background

* show you the tools and calculation methods needed to mix plaster materials.

WHAT IS PLASTER?

Plaster is a building material used for coating walls and ceilings over a variety of background surfaces. It may be smooth or textured, and it may protect the **background surface** from damp or extreme temperatures, or just be decorative. Traditionally, plaster contained lime but gypsum based plasters are now more commonly used and are easier to work with, set a lot faster and provide a high quality smooth finish. Plaster applied to an outside wall or background surface is known as **render**.

As a plasterer, you may work on sites ranging from large construction projects to renovations in clients' homes. It's likely that you'll not only apply plaster materials to walls but also fix plasterboard, lay floor screeds and produce mouldings. This book covers all those techniques and, over time, you may specialise in one or more of them.

However, whatever plastering work you are doing, the quality of the end result is affected by how well you have prepared.

Figure 4.1 A newly plastered wall and ceiling

KEY TERMS

Plaster

– a mixture of dry powder with water that produces a plaster material that is applied to an inside wall when soft and dries as a hard coating.

Background surface

– the surface to which plaster or render will be applied. It would normally be brickwork, blockwork or plasterboard, but may also be concrete, wood, tiles or old plaster.

Render

– the coating on an outside wall. It contains sand, cement and, in some cases, lime or additives to make it weather resistant.

DID YOU KNOW?

You will see that the terms 'plaster' and 'mortar' are often used interchangeably, although they are not quite the same thing. Mortar is usually more like a paste than plaster, and can also refer to lime/sand mortar, as in Practical Tasks 1 and 2 at the end of this chapter.

PREPARING FOR THE PREPARATION OF BACKGROUND SURFACES

Background preparation is one of the most important things to do before plastering begins.

Before applying plaster or render, you must take the time to properly prepare the background. If you don't do this, the plaster will not be smooth and flat, and it may even fall off the wall.

First, the background surface should be checked to make sure that it is **plumb** and flat. Holes and hollows should be filled in, and, if the wall is made of irregular blocks, any projecting blocks must be chiselled off.

The face of the wall should be free from loose dirt and dust, and it must be well dampened to reduce the absorption of moisture from the plaster.

KEY TERMS

Plumb

– the verticality of plasterwork or beads.

Protecting yourself from hazards

Different backgrounds are prepared in different ways. You or your employer should always carry out a risk assessment before work begins (see page 101), so that any issues with health and safety can be identified and dealt with. You must always read the risk assessment and follow its recommendations. If you see anything unsafe, report it to your supervisor at once.

Dust and chemicals are hazardous substances, so ensure you wear the correct personal protective equipment.

Personal protective equipment (PPE)

Always ensure you wear the correct PPE (see Chapter 1). If you are on a construction site, you should wear the PPE provided, such as a hard hat, hi-vis jacket and safety boots. Even on smaller sites, you should wear goggles (to prevent plaster or dust from getting into your eyes) and gloves. Some plasterers prefer to wear thin latex gloves to protect their hands from chemicals but if your employer provides you with thick protective gloves, you must wear them.

Preparing backgrounds can create dust and background coatings often give off fumes, so you might also need to wear respiratory equipment.

Collective safety measures

If you are working at height (for example, rendering a three-storey building) there should be collective safety measures in place, like safety nets or a guardrail with three levels of protection:

* a hand rail at a height of between 1 m and 1.1 m

* a base board between 100 mm and 150 mm

* an intermediate rail.

PRACTICAL TIP

Gloves are not just worn to protect against skin damage from particles. They are vital as many tools cause vibration and gloves will help prevent any long-term damage to your hands.

Figure 4.2 Take care when working at height

Background surface suction

When you are considering a particular background for plastering, you should think about these points:

- Strength: is it in good condition?

- Mechanical key: is it rough or smooth?

- Shrinkage: how will the plaster dry on it?

- Movement: are there joints to allow for the natural movement of the wall?

- Suction: what is it made of?

Different types of background surfaces have different types of **suction**. This is what determines **adhesion** – whether freshly applied plaster will stick to the background. Materials may have low, medium or high suction.

KEY TERMS

Suction

– the porosity or ability to absorb water from an applied material.

Adhesion

– the 'sticking' of a material to the background.

Low suction

Backgrounds like gloss or silk painted surfaces, metal, pre-cast concrete and ceramic tiles are not very **porous**. The fresh plaster is likely to sound hollow when you tap it or even fall off the wall because it will set and cure on top of the surface as there is not enough suction for it to stick. It will need to be treated with a bonding coat or given a key.

Medium suction

Backgrounds like **engineering bricks**, **common bricks**, medium-density blocks, plasterboard and MultiBoard are slightly porous. The plaster will firm up by setting naturally or by evaporation in warmer areas so the background may still need a bonding coat.

High suction

Backgrounds like some types of brick, aircrete blocks and old plaster are porous, so they will suck moisture from the fresh plaster and dry it out too fast. This makes the plaster too difficult to work with as it may be dry before you can make it smooth, or it will crack after you have finished plastering the wall. The suction will need to be controlled with a primer.

Figure 4.3 A low suction background

Figure 4.4 A medium suction background

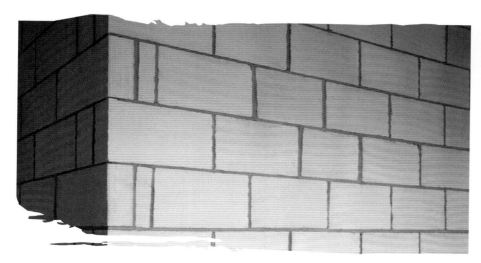

Figure 4.5 A high suction background

Experienced plasterers can tell at once whether a wall is too dry and decide whether additional treatments are needed before the plaster is applied. However, you can do a simple test to find out whether a wall has the right amount of suction. Apply a small patch of plaster to the background, leave it for a few minutes and then test it with your fingers. If it is dry, it is likely to have high suction. If it hasn't dried out at all, it probably has low suction.

PRACTICAL TIP

Another test is to throw a cup of water at the background. If it is absorbed into the wall, it has high suction. If it runs off or stays damp on the wall, the background has low suction.

KEY TERMS

Porous (material)

– something that contains tiny holes that allow water to enter or pass through it.

Engineering bricks

– hard dense bricks of regular size used for carrying heavy loads (e.g. in bridge buildings, heavy foundations, etc.)

Common bricks

– bricks of medium quality used for ordinary walling work where no special face finish is required.

Tools for preparing background surfaces

You will need a variety of tools and equipment in your toolbox when preparing surfaces. Table 4.1 describes the main things you will need.

Hand tool	Description and when it should be used
Bolster chisel 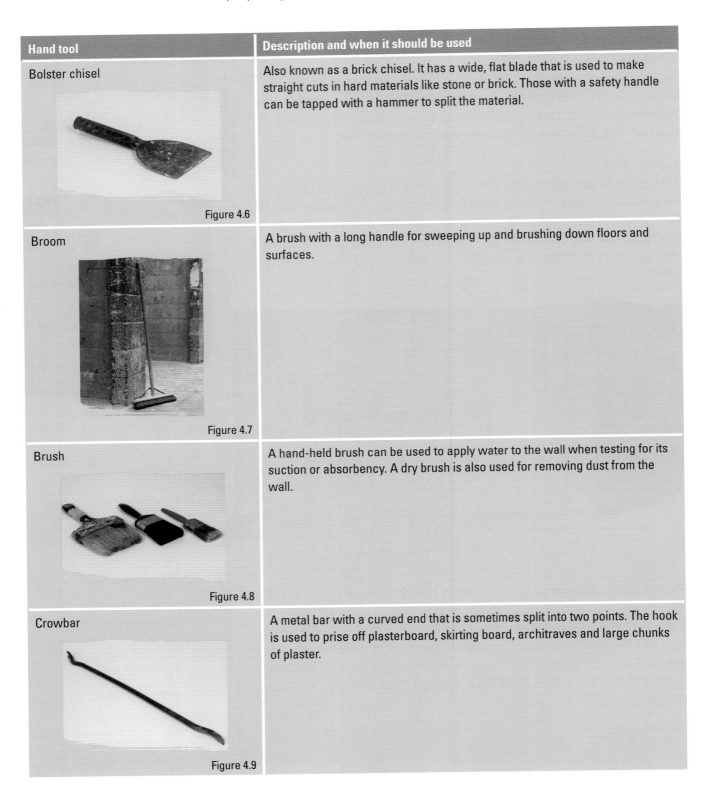 Figure 4.6	Also known as a brick chisel. It has a wide, flat blade that is used to make straight cuts in hard materials like stone or brick. Those with a safety handle can be tapped with a hammer to split the material.
Broom Figure 4.7	A brush with a long handle for sweeping up and brushing down floors and surfaces.
Brush Figure 4.8	A hand-held brush can be used to apply water to the wall when testing for its suction or absorbency. A dry brush is also used for removing dust from the wall.
Crowbar Figure 4.9	A metal bar with a curved end that is sometimes split into two points. The hook is used to prise off plasterboard, skirting board, architraves and large chunks of plaster.

Hand tool	Description and when it should be used
Containers  Figure 4.10	It's useful to have several containers or buckets for different purposes. They can be used to measure or gauge the amount of material you need, as a container for mixing liquid materials (such as bonding and plaster) and for carrying materials or even tools.
Lump hammer  Figure 4.11	Also known as a club hammer. It has a double-faced head, usually made from a heat-treated forged steel. The handle is normally made from wood, usually hickory, or synthetic resin. It may weigh between 0.5 kg and 3 kg.
Shovel  Figure 4.12	This is for mixing materials, loading the cement mixer and for clearing rubble.
Skutch hammer  Figure 4.13	A bricklayer's hammer with interchangeable finishing heads for trimming and tidying bricks and blocks.

Table 4.1 Hand tools and equipment used for preparing background surfaces for plastering

You will also need to use power tools – remember that these must be 110V if you are on a construction site, and 230V if you are in a private building, such as someone's home. Never try to use a power tool if you have not been trained to do so.

Power tool	Description and when it should be used
Hammer drill  Figure 4.14	A drill with a hammer action to help you drill into hard material like brick. The hammer action can be turned on or off, and is usually around 10,000 BPM (blows per minute).
Industrial vacuum  Figure 4.15	Heavy duty commercial vacuums can be used on hard and soft backgrounds, and some remove wet debris. They come in a variety of sizes, suctions and capacities. Some can be attached to drills to automatically remove dust and debris.
Rotary stripper  Figure 4.16	The vibrating abrasive disc removes paint and other surface coatings from background surfaces.
SDS drill 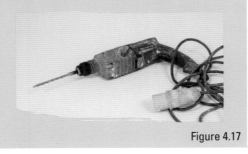 Figure 4.17	This drill is more powerful than other hammer drills and can drill through hard surfaces like stone or concrete. It can take larger drill bits (attachments), including tile removal and chisel bits.

Table 4.2 Power tools used for preparing background surfaces for plastering

CRAVEN COLLEGE

Protecting the work and surrounding area from damage

Throughout your plastering job, you should ensure that your work is not damaged by others on site or adverse weather conditions.

* Plan your work activities to meet the schedule of works.

* Monitor weather conditions for outside work.

* Know the setting times of plaster materials you are using.

* If possible, put a barrier around your work area to stop people from entering it.

* Talk to other trades on site so that everyone knows where and when each other are working.

* If you are outside, cover your work area with a tarpaulin when you leave it. This protects it from rain and damp air.

In Chapter 3, when we looked at sustainability, we learned that it is important to ensure that all construction work has the least possible negative effect on the surrounding area and environment. One of the ways this can be achieved is to keep the site clean and dispose of waste in an environmentally friendly way. This means recycling as much waste as possible.

Following a risk assessment

Your employer should put together a risk assessment for preparing the background surface. It may include some of the risks and control measures shown in Table 4.3.

Source of risk	Type of risk	Control measures
The power tool	• Hair, jewellery or clothing getting tangled in moving parts • Eye injuries from dust or fragments • Hand and wrist injuries from jams and binding • Hand or arm vibration syndrome	• Keep any loose clothing, dangling jewellery or long hair clear of moving parts. • Use suitable eye protection if there is a risk of eye injury. • Only use tools in accordance with the manufacturer's instructions and test them at regular intervals. • Use tools with the lowest vibration levels and minimise the amount of time individuals use the equipment.
The power supply	Electric shock	• Make sure all power feeds comply with European or British Standards and are in good condition. • Where practical only use 110 V electrical tools. • Only use the tools in well-lit and well-ventilated areas.
Chemicals	• Toxic if breathed in or swallowed • Burns • Skin irritation • Eye damage or blindness if chemicals get into eyes	• Use chemicals only in a well-ventilated room. • Follow the manufacturer's instructions and note the hazard symbols used on the label (see below). • Cover up exposed skin on arms and legs. • Use suitable eye, hand and respiratory protection.

Table 4.3 An example of a risk assessment

REED TIP

You'll be using your literacy skills to keep accurate records, e.g. making sure that your van stock is up to date so you don't run out.

Usually risk assessments are graded according to the exposure to hazards. Each of the different tools will also be graded in terms of the injury that they are likely to cause. At one end of the scale tools may produce minor injury, perhaps a blister or a graze. At the other end of the scale they could kill you if something goes wrong.

Chemicals should be clearly labelled to show potential hazards. This includes internationally recognised symbols that show that the chemicals might be explosive, oxidising, highly or extremely flammable, (very) toxic, harmful/irritant, corrosive, or dangerous for the environment. You might see one or more of these symbols on a single product. They used to be orange and black but more recent symbols (pictograms) are red, white and black.

Figure 4.18 Old hazard symbols

You'll see that the harmful symbol is missing. This has been replaced by the exclamation mark pictogram:

Figure 4.19 New hazard symbols (pictograms)

When using power tools and chemicals, you should also follow any relevant laws or official guidelines (see Chapter 1). Your employer should make you aware of any that you need to know about.

PREPARING BACKGROUND SURFACES

Controlling suction

To ensure the plaster sticks and is workable, you must choose the right method of controlling suction before the plaster is applied. Adding a background treatment may seem to add extra time to the job, but it will result in a better finish so will save time in the long run.

There are two main ways of controlling suction: providing a mechanical key or applying a chemical key using bonding agents.

Forming a key to background surfaces

A key provides a rough surface for the plaster to stick to on a low-suction background. This is usually created by raking out brick or blockwork mortar joints to a depth of about 10 mm. Other ways of creating a mechanical key include:

* applying a spatterdash coat (see below)

* cutting zigzags with a craft (Stanley) knife

* hacking the surface with a skutch hammer

* rubbing the surface with a devil float (a float with small nails at one end)

* rubbing the surface with sandpaper or a wire brush

* applying a background treatment containing grit.

Figure 4.20 Raked joint

Figure 4.21 A spatterdash coat provides a mechanical key

Figure 4.22 PVA

Using bonding agents on background surfaces

PVA, EVA and SBR

These are all latex-based adhesives (rubbery glues) and come in a variety of brands and qualities:

* **PVA** (polyvinyl acetate) is one of the most popular bonding agents because it is fairly cheap and is useful for many different purposes.

* **EVA** (ethylene vinyl acetate) is similar to PVA but is for external use. It doesn't contain chlorine so is seen as more environmentally friendly.

* **SBR** (styrene butadiene rubber) is particularly effective in damp or humid areas as it is water-resistant. It may be used on its own or mixed with cement or water.

To apply the bonding agent, follow the instructions on the packaging – you will need to dilute it with water before painting it on the wall and it may be necessary to add sand to provide a rougher texture. Several coats might be needed before the surface is sealed. You will have to wait at least 12 hours before the wall is ready to plaster but you should also ensure it is still tacky when the plaster is applied.

Water

The simplest way to control suction to a low suction background is to spray or paint water over the surface until it runs down the wall, showing that no more water can be absorbed. It is not a true bonding agent but it will increase suction enough for plaster to be applied. It is only a short-term solution and it is better to use a chemical bonding agent.

Primers and stabilisers

Figure 4.23 SBR

These are specially formulated to form a bond for plaster. They are usually water-based polymers containing fine sand or **aggregate**, which gives them a gritty texture that the plaster attaches to. Various brands are available, and are applied with a brush or a roller, or can be sprayed. They often dry more quickly than PVA but you should plaster when it is still tacky.

KEY TERMS

Aggregate

– the coarse material that is the bulk of a mix.

Figure 4.24 A common brand of primer

Methods of cleaning down background surfaces

Oil, grease, films, salt, dirt, dust and other loose material can interfere with bonding, so they need to be cleaned off. Methods depend on how dirty the background surface is, where it is and what it is. They include:

* **brushing down** using a brush or broom

* **scraping** old plaster and dirt off the background surface with a scraping tool

* blowing with (oil-free) **compressed air** but ensure the resulting dust is collected

* **wetting** with a damp cloth or sponge

* **vacuum cleaning** to suck off the loose dust

* **wet sand-blasting** very dirty surfaces

* **water jetting** very dirty external surfaces.

Do not use solvents to remove films formed by curing compounds.

Figure 4.25 Scraping off a surface

Figure 4.26 Water jetting

Applying spatterdash or stipple coat on background surfaces

Spatterdash

Spatterdash is a type of mechanical key that provides a surface for the plaster to stick to. It is a wet mix (slurry) made of:

* cement and sand mixed to a proportion of 1:1.5 or 1:3, or

* equal parts sand, cement and PVA or SBR

Figure 4.27 Spatterdash

Figure 4.28 Stipple coat

with added water. You can buy it as a dry mixture and add the water when you're ready to use it. The spatterdash is traditionally thrown, or spattered, against the surface, to a thickness of 3–5mm, using a dashing trowel, a block brush or spraying it on with a machine. While you are new to plastering, you will get a better result by trowelling it onto the wall, as shown in Practical Task 3 on page 122.

The spatterdash should completely cover the surface and form a rough layer. It should then be allowed to harden.

Stipple coat

A stipple coat slurry is similar to a spatterdash coat but is mixed from one part of cement and one and a half parts sharp sand before adding water and a bonding agent like SBR.

Instead of being thrown at the surface, the mixture is pushed into the surface with a coarse brush and then dabbed with a refilled brush. The resulting coarse finish should be protected from rapid drying out for a day and then left to harden for another one to two days.

Using dubbing out coats on background surfaces

If there are hollows or deep joints in the background surface, you will need to apply a **dubbing out** coat to provide a level flat surface. A **scratch coat** of undercoat plaster can be applied over the dubbing coat, keyed and left to dry for 24 hours before applying a second coat of plaster.

It should be applied as layers of up to 15mm, which are left to dry before the next layer is applied. Once it is set (usually after a few days), you can add the key or scratch coat.

KEY TERMS

Dubbing out

– a mixture used to fill large hollows or voids before applying a scratch coat.

Scratch coat

– the first coat of plaster materials applied, to control suction, straighten and even out walls and provide a mechanical key for the next coat.

PRACTICAL TIP

For external work and repairing older buildings, lime plasters may be specified by the client as these will help to slow down the setting process and reduce shrinkage.

DID YOU KNOW?

Animal hair has traditionally been used as a binding agent within base coats of external renders and within internal plasterwork. It helps to add flexibility to the coat and reduces shrinkage. Once hair is added to a plaster, it should be used immediately.

Common faults caused by ineffective background surface treatments

If backgrounds are not prepared properly, your plastering will be a waste of time. After you have plastered, there may be problems like movement. This can occur because of structural movement due to the design of the wall or background and may result in the plaster cracking. To avoid this, EML (expanded metal lathing – see page 156) may be applied to weak areas, for example where brick and timber meet.

Another common issue is the plaster cracking because a high suction background was not corrected. You should always take the time to treat the suction using suitable methods.

Maintaining a clear work area

You must keep your work area clean and tidy at all times. There are several reasons for this:

* A messy work area increases the risk of accidents.

* Leaving tools and materials lying around looks unprofessional to the client.

* Discarded tools and waste can make it difficult for other trades on site to work around you.

Clear up after yourself regularly and at the end of the day. Always brush the area with a broom or dustpan and brush when you have created dust, or use an industrial vacuum cleaner if disturbing a lot of dust is likely to cause respiratory issues.

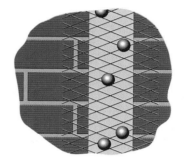

Figure 4.29 EML applied over a joint to allow for movement

Dealing with waste

The Building Act (1984) states that it is the construction industry's responsibility to prevent and control waste. It should also make sure that resources are not wasted unnecessarily.

Part H of the Building Regulations covers disposal of all types of building materials.

The drive towards sustainable and secure buildings also aims to control waste and to protect the environment. You could refer back to Chapter 3 to refresh your memory about sustainability.

Collect and remove any rubble, bricks, pallets and old or dried plaster. Recycle or dispose of it according to your site rules.

Control measures

Control measures may include:

* wetting down dust before removing it

* placing drop boards (for example, old plasterboard) on the floor or ground under the wall to catch any material that falls

* storing bags of plaster in dry areas, away from the floor and walls, which may be damp

* clearing up any spillages as soon as you notice them, even if they weren't your fault

* storing or plugging in cables safely to help prevent trips

* reporting any defects that you see to equipment or work.

Methods of removing existing plaster

Old plaster can be difficult to remove, especially if it was applied to brickwork using a bonding adhesive.

There are several ways to take off old plaster, depending on how difficult the plaster is to remove.

* Hit the wall with the head of a claw hammer to loosen the plaster and then use the claw to pull the plaster off. This works best with timber backgrounds.

* If the plaster is already fairly loose, place the blade of a bolster chisel behind it and hit it with a lump hammer.

* For larger expanses, you can push a spade behind the plaster – though this method requires quite a bit of strength.

* For tough plaster and Artex, use an SDS drill with rotary stop set to the hammer-only setting and with a 40 mm tile removal or chisel bit.

PRACTICAL TIP

When removing old plaster with an SDS drill, place the chisel bit at a 45° angle between the brickwork (or other background surface) and the plaster, pushing the drill away from you along the wall. Using it straight on, at an angle of 90°, will loosen soft bricks in the wall. Some plasterers prefer to start from the bottom and work up to about a metre before removing the loose plaster.

Figure 4.30 Using a hammer and bolster chisel to remove old plaster from a wall

Let smaller loose chunks of plaster fall to the floor but once you've loosened larger pieces, stop the drill and pull them off with your hands. This gives you more control over their removal, and reduces the chance of heavy pieces falling onto your foot or smashing when they hit the ground.

Clearing up

Removing plaster is a very dusty job that produces a lot of debris. Allow time to clean up properly – leaving rubbish around is dangerous and looks unprofessional.

* Ensure that doors and windows are closed.

* Make sure that nobody can enter the working area without PPE.

* Wear a dust mask, safety glasses, safety gloves and safety shoes.

* Always clean your tools when you have finished using them.

KNOW HOW TO PREPARE FOR AND MIX PLASTER MATERIALS

Procedures for the material mixing area

Before you start mixing plaster materials, make sure that:

* the area is well ventilated, for example with windows and doors open

* the floor is level and the container is large enough for the mix

* you wear appropriate PPE, such as a dust mask to avoid breathing in dust

* there is nothing that can contaminate the mix, for example nearby work that is creating dust

* your mixing equipment is clean

* there is enough space for you to mix the quantity you need.

Maintenance and manufacturers' instructions

It is much quicker to mix plaster materials with the appropriate tools rather than by hand. However, these should be treated like any other power tool and only used after training and familiarising yourself with the particular make or brand.

Table 4.4 outlines the main tools and machinery you will use to mix plaster materials, and how to look after them.

Tool / machinery	Description	Cleaning and care
Cement mixer  Figure 4.31	A self-contained mixer is useful for larger jobs. Portable cement mixers have a capacity of 60 to 150 litres. They are powered by electricity (110V or 230V) or petrol.	• Clean the mixer by running it with water or water and brick or gravel inside. • Ensure any water is drained after use. • Ensure the mixer is switched off after use.
Mortar mill/pan mixer  Figure 4.32	This grinds and mixes powders and suspensions and is regarded as more efficient than a cement mixer, especially for the mixing of lime plaster and crushing materials to a lump-free plaster mix. Available in a variety of sizes.	• Always ensure that only materials specified by the manufacturer are placed in the mill. • Clean the mill thoroughly after use
Whisk/paddle mixer  Figure 4.33	This is a power tool with a long stem that has a whisk (paddle) attachment at the end, for mixing viscous materials like adhesive and plaster. It will usually have a two-speed motor for mixing both thin and thick materials, as well as variable speed control. Many different types of attachment are available.	• Clean the whisk by switching it on in a bucket of clean water. • Do not let the plaster on it dry or it will be difficult to remove.

Table 4.4 Powered tools and machinery for mixing plaster materials

Cleaning

The majority of power tools are designed to be able to be used for long periods of time without any real maintenance. However, manufacturers do recommend that power tools are checked and cleaned on a regular basis. This is because over time they could lose their efficiency. Using dirty tools will also contaminate your mix, which could affect the setting time of the plaster or make the new batch look different from the previous batch.

Another reason to clean your tool is that, if you are using tools that produce dust or you are working in dusty areas, the dust will be attracted to the motor of the power tool and this could damage it and make it overheat.

PRACTICAL TIP

When using a whisk or paddle mixer, keep an eye on the cable and make sure it is behind you when mixing.

PRACTICAL TIP

Obviously power tools cannot be cleaned with water but you can remove dust from an SDS drill by spraying it with a can of compressed air, making sure you spray the vents and areas around the control buttons. You can also use special tool-cleaning wipes to remove any remaining dirt.

PRACTICAL TIP

Don't be tempted to unscrew and take to pieces any power tools or have anyone other than an authorised repair person carry out the work. Not only are you putting yourself at risk due to a poor repair, but it will also mean that any manufacturer's guarantee on the tool will be invalid.

Manufacturers' guidance

Each manufacturer of tools provides a data sheet or instruction manual, which is often available online. For the majority of situations this will:

* give you information on how to clean and maintain the power tool yourself. This will tell you what you are able to do on your own

* inform you of situations when you should get an authorised repairer to deal with a problem.

Checking for damage

You should regularly check your power tools. This means checking for damage on the tool itself and clearing away any dust from ventilation slots. You also need to check the tool's lead and plug for any damage.

Storage

Manufacturers will also recommend how to store the power tools. Mostly this means putting them back into their case and making sure that they are not stored in wet or cold conditions.

Procedures for operating relevant power tools and supplies

Portable power tools and larger machinery like concrete mixers can speed up many routine tasks but you must be aware of the risks connected with using them. According to the Health and Safety Executive, a quarter of all reportable electrical accidents involve portable power equipment.

Power tools should only be used by those who are competent, so before using them you should have training and know exactly how to operate the tools. If you do not then you will be putting yourself and others who are working near you at risk.

You should always make sure that all necessary precautions have been taken before using a power tool. For example, you should do the following:

* Make yourself familiar with the way the tool works and have practised using it in controlled conditions.

* Refer to the manufacturer's instructions before using the tool.

* Give the tool a look over to make sure it is safe to use and know what to do if there is something wrong with it.

* Check the plug and cable for any damage.

* Check the tool is clean – if not, clean it.

* Check that if you are using a mains power source, it is a safe voltage.

* Find out if the power tool should have guards and, if so, check that these are correctly fitted.

* Check any blades, drills, cutters or bits before using them. Make sure they are not worn or damaged.

* Make sure you are wearing the right PPE and RPE. This could include ear defenders, gloves, a dust mask and eye protection.

* Make sure you are not wearing any loose clothing as this could easily become caught in the tool and injure you, or make the tool overheat.

DID YOU KNOW?

On a construction site, you should always use a 110 V power source. In domestic settings, such as customers' homes, the power source is 230 V.

When using a power tool, keep safety in mind at all times. Table 4.5 shows some points you need to remember.

Always...	Never...
...make sure the tool is switched off before plugging it into the electrical supply.	...try to use a 110 V drill on a 230 V supply.
...fully unravel the extension lead.	...trail the lead in oil or water.
...isolate tools from the supply before cleaning or repairing them.	...lift or lower tools by the cable.
...report any faults or damage to your supervisor.	...mess around with power tools.
...stop work if you think anything is wrong with the tool or if it is not having the effect you expect.	...use power tools when they are wet, in wet conditions or outside when it is raining.
	...attempt to make any repairs yourself – report damage or faults to your supervisor and don't use the tool.
	...change a blade, bit, tip or drill without first disconnecting the tool from the power supply.

Table 4.5 Points to remember when using a power tool

Methods of mixing plaster materials

The method you will use to mix plaster will depend on how much plaster you need, and where you are working. Table 4.6 shows the three main ways of mixing plaster.

Method	When to use this method
With a shovel or rake	• When mixing smaller quantities, for example when repairing a wall • When no power source is available
With a whisk or mixing drill	• When mixing lightweight plaster in a trough • When mixing one coat plaster or finishing plaster in a drum or bucket • When you have a reliable power source
With a cement mixer	• When mixing large quantities • When mixing external rendering sand and cement • When you have enough space and a reliable power source

Table 4.6 Different methods of mixing plaster materials

PRACTICAL TIP

If you are mixing in a bucket, the plaster will heat up in the enclosed space and may set too early. Try adding a little extra water to stop this from happening.

To make plaster of the correct consistency, you need to use its constituent materials in the correct proportions. Materials are measured by weight or volume, using the same method for different materials and batches. You can use any suitable container to **gauge** the amount you need, such as a bucket, bag or wheelbarrow, or you can use a **gauge box**. Pour in the materials, level them off and remove the gauge box to leave the correct amounts behind.

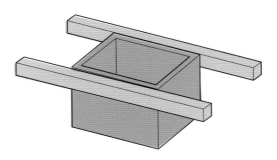

Figure 4.34 A gauge box

Types of additives

Chapter 5 describes different types of plaster in more detail. However, before you start to mix the plaster, it is useful to know about the additives that may be added into it.

Cement is a powder that goes through a chemical change or reaction when mixed with water. It becomes like an adhesive paste and then hardens. Its main purpose is to bind bulkier materials like sand and aggregates within the plaster mixture.

Without additives, cement can make plaster shrink and crack as it dries. Too much water in the mix will make the plaster weak. In hot weather the mix can dry too quickly and in cold weather it can dry too slowly or be damaged by frost. Standard mixes are not waterproof and most mixes will develop surface pores as air bubbles escape while the mixture dries, causing weakness and allowing water penetration.

Additives are designed to make working with plaster easier and to produce better results. Additives may have been added to the plaster when it was manufactured or you may need to add them yourself.

Table 4.7 describes some of the additives you are most likely to use.

Additive	Description
Frostproofer/ accelerator	This is a liquid additive formulated to accelerate setting and hardening times of plaster, concretes, screeds and rendering, so that it is not affected by frost while it is setting. It can be effective in sub-zero temperatures and can also be used in normal temperatures where a rapid set is required. Do not use this additive with lime cement. Add frostproofer to water at a rate of 1.7 litres to every 50 kg of cement .
Plasticiser	This is added to the water to make the plaster smoother and easier to work with. It may be added instead of lime, as it has its advantages but is cheaper and maintains the strength of the mix. Less water needs to be added to the mix, which helps to prevent cracking and shrinkage. It gives the mix additional strength and flexibility. It also often delays the setting time, which makes it suitable for use over large areas. The plasticiser liquid is mixed with water at a ratio of 1:100.
Waterproofer	This plasticises the mix by preventing water from penetrating cement plasters without acting as a vapour barrier. It also reduces suction when it is used in cement-based undercoats. The waterproofer liquid is mixed with water at a ratio of 1:30 and added to the cement/aggregate mix to achieve the required consistency.
Coloured pigments and dyes	These change the colour of the plaster but should have no other effect on it. They usually come in the form of granules which are combined with the dry mix before the water is added. Depending on the depth of colour required, the pigment should be added at a ratio of 2:100 to 10:100. Note that the same quantity needs to be used in different batches to prevent variations in the colour of the plaster.
Expanded perlite	This is a glassy volcanic mineral that contains a small amount of water. When it has been crushed and superheated, this water turns to steam, which helps to form a fine substance that expands its volume up to 20 times. It is fireproof, soundproof and insulating, with a warmer surface than standard gypsum plasters that prevents condensation. It is used in lightweight plasters.
Vermiculite	This is produced in a similar way to expanded perlite, and has similar properties. It has a wide variety of uses but it is used as a lightweight aggregate in plaster applied either by hand or as a spray to improve coverage, ease of handling, adhesion to background surfaces, fire resistance, and resistance to chipping, cracking and shrinkage.
Fibre-reinforced gypsum	This ready-mixed gypsum contains glass fibre, which helps to prevent cracks in the plaster caused by shrinkage.

Table 4.7 Types of plaster additives

Hazards and control measures when mixing plaster materials

Dry powder can get into the air, and the materials that go into the plaster mix often contain ingredients that irritate the skin. Ensure that you are wearing the correct PPE and respiratory protective equipment (RPE), such as goggles, gloves and dust mask before starting to mix.

Other control measures include:

* Always use clean water to avoid the mixture being contaminated.

* Only mix in well-ventilated areas.

* Don't use plaster that has passed the use-by date stamped on its bag.

KEY TERMS

Plasticiser

– an additive used to make plaster more workable.

Perlite

– an aggregate formed from volcanic rock, found in lightweight plasters with good fire-resistance qualities.

* Put the water into the bucket first then add the powder slowly, bit by bit, letting it soak in before adding more.

* Gauge and mix the correct amount.

* Clean all equipment after use.

* Only use 110V power sources for your power tools if you are on a construction site.

* Keep the area tidy at all times.

Under- and over-mixing plaster materials

As soon as the powder and water meet, a chemical reaction begins that starts the setting process. If mixed correctly, the plaster should last for around 15 minutes, or 20 to 30 minutes in colder conditions, before starting to set.

The longer you mix the plaster, the stronger it will be. However, after mixing it for 10 minutes or so (depending on the mix, mixer, temperature and other factors) it will start to become weaker again. The ideal time you need to take when mixing comes with experience and will not be the same for every batch.

Under-mixing will result in a plaster that has a lumpy or watery consistency, caused by the powder and water separating. It will not set properly and will be difficult to apply to a wall or ceiling.

Over-mixing – that is, mixing for too long or at too fast a speed setting – will create heat that makes the plaster set more quickly. You might find that you can't spread it on a wall.

Mixing too large a batch could have a similar result – it could start to set before you are ready to pour it.

Protecting the work and disposing of waste

Think about other people on site – for example, other trades or the householder if you are working in a domestic setting. You don't want them to damage or spoil your work, and you also don't want anyone to be hurt. In particular:

* Keep tools and materials tidy and out of the way of anyone who will be in the area.

* Don't leave mixed plaster materials to set if you are not going to use them.

* Be careful of splashing painted walls and the dust caused by mixing plaster indoors.

Disposing of gypsum waste

In Chapter 3, when we looked at sustainability, we learned that it is important to ensure that all construction work has the least possible negative effect on the surrounding area. One of the ways of reducing its negative impacts is to keep the site clean and dispose of waste in an environmentally friendly way. This means recycling as much waste as possible.

It is illegal to send plasterboard and gypsum to landfill, mixed in with other waste. This is because gypsum, when mixed with biodegradable waste, can produce hydrogen sulphide gas in landfill. This gas is not only toxic but also smells unpleasant. The Environment Agency recommends other steps to take instead:

* Separate gypsum-based material and plasterboard from other wastes on site so it can either be recycled, reused or disposed of properly at landfill.

* Do not deliberately mix gypsum or plasterboard waste with other waste for landfill.

* Enquire whether your company's waste management contractor will provide separate skips for the waste or sort it for you.

In any case, all waste, no matter what it is, should be sorted and treated before it is sent to landfill.

CASE STUDY

South Tyneside Homes

South Tyneside Council's
Housing Company

You're learning all the time

Gary Kirsop, Head of Property Services at South Tyneside Homes says:

'It's important to take your time. It's not about speed, it's about understanding the importance of health and safety so you have a good grounding on live sites straight away. Make sure you do know the basics in college because the hazards on a construction site are real and bigger. Listen to your tutors and to check that you're doing it right.

And when you're on your apprenticeship, listen to your mentors and team leaders. Don't talk over them and say, "I know better" – everybody is learning every day. Listening is a vital part of being a good communicator, and the best way to learn is from tradespeople who are more experienced than you. If you're not sure how to do something, it is always better to go back and ask for help.

Remember that you'll be using real materials, working for real clients, real people, with real needs, who are spending real money.'

1. MIX LIME/SAND MORTAR USING A PAN MIXER

OBJECTIVE

To mix lime/sand mortar to a good consistency ready for application to a solid brick wall.

INTRODUCTION

As a plasterer it's essential you understand how to mix materials and what quantities of materials are required. In this task, you will be using lime/sand mortar, which takes longer to set than gypsum plaster, but as you get more experience you will be able to work more quickly.

TOOLS AND EQUIPMENT

Plastering trowel	Spot board
Handboard/hawk	Buckets or gauge box
Gauger	Pan mixer

PPE

Ensure you select PPE appropriate to the job and site where you are working. Refer to the PPE section of Chapter 1.

STEP 1 Check the pan mixer is as clean as you can make it and in good working order.

Figure 4.35 A pan mixer

STEP 2 Use a gauge box or buckets to measure and mix the correct proportion of lime and sand. This may be 1 part lime to 3 parts sand but your tutor might ask you to mix a different ratio.

STEP 3 Add 1 bucket of clean water to the pan mixer then 6 even shovels of lime/sand mortar – about half your final mix. Switch on the pan mixer straight away and leave it to mix for a few minutes.

PRACTICAL TIP

Always add the mortar to the water, not the other way around.

Figure 4.36 Letting out the mortar

STEP 4 Add another 6 even shovels of lime/sand mortar and allow the pan mixer to mix it for a few more minutes. Check the consistency by taking a small amount on the gauger. The material should be soft and pliable and not too dry.

Figure 4.37 Mortar mixed to the correct consistency

PRACTICAL TIP

Remember that under-mixing can make the mixture lumpy and difficult to apply to a wall, so always check the mix is right before moving it onto the spot board.

STEP 5 Empty the mix into a bucket or fill three-quarters of a wheelbarrow. From here, place enough on the spot board to work with, without having to keep refilling it.

Figure 4.38 The mix on the spot board

STEP 6 Clean the pan mixer and all the tools and equipment you have used. The lime content slows the set so there is enough time to clean your equipment before it goes off.

STEP 7 Ensure all electrical equipment is switched off before using the mortar.

2. APPLY A SCRATCH COAT OF LIME/SAND MORTAR

OBJECTIVE

To apply a scratch coat of lime/sand mortar to a small wall area and to provide a mechanical key.

INTRODUCTION

This task will introduce you to using a handboard and plastering trowel and help you to become familiar with tool manipulation by transferring mortar onto the trowel. You will use a slow-setting material to cover a small area of solid brick wall and then provide a mechanical key using a scratcher.

TOOLS AND EQUIPMENT

Plastering trowel	Spot board
Handboard/hawk	Hop-up
Flat brush	Bucket of mixed lime/sand mortar
Scratcher	

PPE

Ensure you select PPE appropriate to the job and site conditions where you are working. Refer to the PPE section of Chapter 1.

Figure 4.39 A plastering trowel, handboard and scratcher

STEP 1 Work out how much mortar you need. As a general rule, a bucket of mixed mortar will cover approximately 1 square metre of wall area. Ensure your spot board is clean and damp it with a little water. Load your material onto the board, to a maximum of 3 buckets of mortar.

STEP 2 Practise taking the mortar from the handboard to the trowel.

Figure 4.40 Moving mortar from the handboard to the trowel

Before work can start, you need to check that the background is suitable:

- Does any old plaster need to be removed?
- Do the joints need raking out?
- How straight is the wall?
- What is the suction like? Check it by wetting the background. If the water is immediately absorbed (disappears into the wall) it is a high suction background and you need to treat it with PVA or a similar base.

Make sure the wall is properly prepared before beginning to plaster it.

Make sure all areas of the wall are covered evenly. Go over any hollows or bumps and tighten in (smooth) the mortar.

STEP 5 When you have covered the required area, use a scratcher to form a key. Start from the top and use wavy horizontal strokes close together across the wall.

Figure 4.42 Forming a key with a scratcher

STEP 3 Starting at head height on the right-hand side of the wall, apply an even coat of mortar.

To apply an undercoat, start from the right. For skimming finishing plaster, start on the left (or vice versa if you are left-handed).

Figure 4.41 Applying the mortar

STEP 6 Once you have formed the key, clean the floor area of any materials that have dropped onto it and wash all your tools with clean water.

Figure 4.43 The completed keyed surface

STEP 4 Apply the mortar above head height by using a small hop-up to reach the top of the wall.

3. APPLY A SPATTERDASH OR STIPPLE COAT TO A SOLID BRICKWORK BACKGROUND

OBJECTIVE

To apply a slurry or spatterdash coat to a solid brick, stone or block wall.

INTRODUCTION

Spatterdash coats are applied to walls when a mechanical key is required before applying a scratch coat. It is made up of a slurry of materials and applied with a brush or dashing trowel, or thrown at the wall, depending on the type of background and the consistency of the materials.

You will be working on the wall that you prepared in Practical Task 2.

TOOLS AND EQUIPMENT

Plastering trowel	Hand brush
Handboard/hawk	Spot board
Gauger	Hop-up
Dashing trowel	Buckets
Flat brush	Cement mixer

PPE

Ensure you select PPE appropriate to the job and site conditions where you are working. Refer to the PPE section of Chapter 1.

Figure 4.44 Stipple brush

STEP 1 Set up your mixing area, with PVA/SBR cement, sand and clean water to hand.

STEP 2 Mix the cement and sand to a proportion of 1:3, or equal parts sand, cement and PVA/SBR.

Figure 4.45 Mixing the slurry

PRACTICAL TIP

Always use a bucket under your arm, half-filled with the slurry so it is not too heavy. This catches any materials that fall off the wall so they are not wasted.

STEP 5 When you have covered the wall, clean all your tools and equipment and leave the wall to set for a minimum of 24 hours before applying the next coat.

Figure 4.47 The finished spatterdash wall

PRACTICAL TIP

Wear safety glasses when mixing and applying the slurry mix.

STEP 3 Damp the wall down using a flat brush and remove any old plaster, cement or mortar.

STEP 4 Apply the slurry to the wall, starting at the top. If you are using heavy material, use the dashing trowel, keeping it at 90° to the wall for an even surface. If you are using a lighter mix, use the brush. If you are brushing the slurry onto the wall, make sure you cover the whole area, leaving no gaps.

Figure 4.46 Applying the slurry

TEST YOURSELF

1. Which of these is an example of a high suction background?

    a. Plasterboard

    b. Pre-cast concrete

    c. Brickwork

    d. Ceramic tiles

2. What is porous material?

    a. Material that allows water to pass through it

    b. Very heavy material

    c. Material that does not allow material to pass through it

    d. Material that causes skin burns

3. What is a mechanical key?

    a. A chemical added to a background to make the plaster stick

    b. A rough surface for the plaster to grip onto

    c. A type of hand tool used for mixing plaster

    d. A high suction background surface

4. What happens if you over-mix plaster?

    a. It will go lumpy

    b. It will never set

    c. It will set too quickly

    d. It will separate out into water and powder

5. Which of these is not a recommended method of cleaning background surfaces before plastering?

    a. Brushing down

    b. Wetting with a damp cloth or sponge

    c. Vacuum cleaning to suck off the loose dust

    d. Applying a solvent

6. What is the main reason for adding a plasticiser to the mortar mix?

    a. It changes its colour

    b. It makes the plaster smoother and easier to work with

    c. It makes it set more quickly

    d. It is fireproof

7. What might you use an SDS drill for?

    a. Removing old plaster

    b. Mixing plaster

    c. Applying bonding

    d. Making the background material more porous

8. Why would you apply a dubbing out coat?

    a. To make the plaster stick to the background

    b. To make the surface rougher

    c. To fill in holes in an uneven surface

    d. To increase suction

9. What is a gauge box?

    a. A bottomless box used for measuring out materials

    b. The case that is supplied with a gauge

    c. A box for measuring out liquids

    d. A container to mix plaster in

10. What should you do if the power tool you are about to use is damaged?

    a. Mend it yourself and get on with the job

    b. Use it carefully, keeping alert for any problems

    c. Give it to a more experienced colleague to use

    d. Do not use it and report it to your supervisor

Unit CSA–L1Occ17
APPLY PLASTER MATERIALS TO INTERNAL BACKGROUNDS

LEARNING OUTCOMES

LO1/2: Know how to and be able to prepare for applying plaster materials to internal backgrounds

LO3/4: Know how to and be able to apply materials to internal backgrounds

LO5/6: Know how to and be able to apply plaster to internal solid backgrounds

INTRODUCTION

The aims of this chapter are to:

* show you how to recognise the tools and equipment needed to apply plaster materials

* show you when to use different types of plaster materials

* help you to become familiar with applying plaster, plasterboard and metal components

* show you how to apply one, two and three coats of plaster.

KEY TERMS

Plasterboard

– a panel of gypsum plaster that has been pressed between two thick sheets of paper. One side of the plasterboard is used for dry lining and the other side for plastering.

Dry lining

– applying plasterboard to background surfaces like brickwork, timber or metal to provide a flat surface, to secure insulation or to provide the walls with waterproof, fireproof or soundproof properties.

PREPARING TO APPLY PLASTER MATERIALS TO INTERNAL BACKGROUNDS

As a plasterer, you will use both plaster and plasterboard in your work, so it is important to know how to apply both. **Plasterboard** has a variety of applications, from **dry lining** walls and ceilings to forming new internal walls and providing moisture, sound, fire and thermal resistance. It provides a flat surface for plastering or painting.

When you are ready to start plastering and fixing plasterboard it is important that you begin by selecting the right materials and tools. Accurate positioning, following either full-size or scaled drawings, is also very important.

Whenever you are dealing with plaster or plasterboard you also need to be aware of health and safety risks and wear the appropriate PPE.

Hazards, health and safety, and risk assessment

You or your employer should carry out a risk assessment before work begins.

The main hazards when working with plasterboard are associated with manual handling, and using hand tools when you cut and fix the board.

The main hazards when working with plaster are from the chemicals in the mix.

Other hazards arise if, when fixing plasterboard or applying plaster, you need to work from height or need to move the materials. Here are some basic controls to lessen these risks.

* Always make sure you are wearing the correct PPE (such as gloves, safety shoes and goggles) before you start work.

PRACTICAL TIP

Although your manager will probably be the one to write the risk assessment, you need to be familiar with the types of hazards and controls it will cover. You might also identify risks that your manager wasn't aware of.

* If you are on a large site, a forklift truck or pallet truck may help you to avoid carrying heavy materials, large pallets and waste.

* Transport small amounts of mixed plaster and mortar by wheelbarrow.

* When plastering, inspect mixers, barrows and shovels regularly for damage or defects and take them out of use for repair if faults are discovered.

* Fence off your work area, especially if you are working at height.

* If possible, use warning signs at all access points to the work area where work is being carried out above ground level.

You should make sure that you refresh your memory about general health and safety precautions by referring back to Chapter 1.

Hazards when working with plaster

Plaster materials that may be hazardous include:

* adhesives

* cement and other gypsum-based products

* joint fillers

* lime

* mineralised methylated spirit

* polyester resins

* PVA

* shellac.

These can all damage your skin or cause illness if they are breathed in. To reduce the risks:

* Store all materials in an approved, enclosed store or location.

* Refer to the manufacturers' data sheets for each substance.

* Use additional PPE, as needed, when mixing and applying plaster – such as overalls and a dust mask.

* Maintain good standards of personal hygiene by washing your hands after contact with the materials and before eating, drinking and smoking.

> **DID YOU KNOW?**
>
> The Health and Safety Executive (HSE) has produced guidelines for small plastering companies on drawing up risk assessments. Go to www.hse.gov.uk and type in *risk case studies plasterers.*

> **PRACTICAL TIP**
>
> On a domestic site, you need to agree with the householder where to store your materials. If there isn't much space, it may have to be the corner of a room, as long as it is dry and will be undisturbed. If you leave your materials in the house overnight, make sure the householder is aware of the hazards of handling the materials.

Hazards when working with plasterboard

Manual handling

Gypsum-based sheet materials are heavy. A standard sheet of 1,200 mm × 2,400 mm plasterboard that is 12.5 mm thick will weigh around 23.5 kg. This means a pallet of 72 sheets will weigh 1.6 tonnes *(source*: wrap.org.uk). You can therefore injure your back or strain your muscles if you do not consider the risks of manual handling.

Where possible, you should use mechanical means to transport boards, such as a forklift truck. Only operate this machinery if you have been trained to do so. Where this is not practical, for example if you are working in someone's home, you will have to carry the boards yourself. Always carry plasterboard on edge, not flat, to avoid putting unsupported pressure on the board's face. If possible, get someone to help you move it.

Bend your knees, **not** your back

Always lift with a **straight back**

The sheet needs to be lifted only a little way off the floor

Figure 5.1 How to carry plasterboard safely

There are also risks when installing the plasterboard, especially to ceilings. You may strain your neck or injure your arms by holding boards above your head. You may also need to hold your wrists at an awkward angle when drilling holes to fix ceiling boards. Mechanical aids such as ceiling lifts and adjustable props will reduce the pressure when working with boards above head height. Many workers prefer to use these aids as they generally make the job easier.

Using tools

You will use a craft or trimming knife (often known as a board or Stanley knife) to cut plasterboard to size. You will use a hammer or drill to fix it in place. Although these are fairly simple tools, hazards may arise as a result of them being misused or poorly maintained.

It is your responsibility and that of your employer to make sure that any tools you are using are in good condition and safe to be used. The main causes of injury are:

* not using the right tool for the job

* not having been trained to use the tool in a safe way

* deliberately using the wrong tool

* failing to maintain the tool

* not wearing the right PPE.

Tools need to be regularly inspected and some will need routine maintenance to stop them from being potential hazards. Cutting tools, such as craft or trimming knives, will often need their blades replacing. Wooden handles must be free of cracks and splinters.

If a tool cannot be properly repaired it should be replaced.

When you are carrying out a risk assessment remember that poorly maintained hand tools are dangerous. Tradespeople rarely use poor quality tools, partly because you are more likely to injure yourself with these types of tools.

Using information sources to position plasterboard and plastering

As we saw in Chapter 2, various types of documentation help you to plan your work and position the plasterboard or plastering correctly.

Working drawings are either full size or scaled, accurate illustrations of the final finish that is required. From these working drawings you will be able to mark off the plasterboard to the required dimensions to allow you to make the necessary cuts. Equally important is where you put the fixings as you may accidentally drill into cables or pipes if you have not checked their position beforehand.

It is important to follow any supplied documentation to ensure that your work complies with Building Regulations and other requirements of the project. For example, the technical drawings may specify particular types of plasterboard to ensure structural stability, acoustic performance and fire resistance.

You should also refer to the plaster or plasterboard manufacturer's technical data sheets to make sure it's the correct material for the area in which it will be applied.

Using the specification to select appropriate plaster and plasterboard

Specifications list the sizes and type of material required, along with the quality of the finish. They will list:

- whether any insulation or studwork needs to be in place first
- the thickness of the plasterboard required
- any special properties required for the plaster or plasterboard, for example soundproofing, fire resistance or water resistance
- recommended brands of plaster or plasterboard, if the architect has a preference.

Outline Specification

3.0	**EXTERNAL WALLS**	
3.1	**Internal Wall Lining to existing outbuilding walls, for new home office:** 62.5mm (total thickness) Kingspan Kooltherm K18, or similar, insulated plasterboard with skim finish, on Tyvek Housewrap, or similar, breather membrane, fixed to 50mm x 25mm vertical timber battens, at 600mm centres (and horizontally at floor and ceiling level), on strips of DPC (between battens and masonry), on existing half-brick external wall. Insulated wall linings to extend into the reveals to minimise cold bridging. The above proposed construction will provide a u-value of 0.3 W/m^2K. (Building Regulations AD Part L minimum requirement: u-value of upgraded wall construction not to exceed 0.30 W/m^2K.)	
3.2	**New external wall between home office and utility room:** Studwork wall to be formed with 97x47mm C16 studs at 400mm centres, with 75mm Kingspan Kooltherm K12 insulation between studs. Internal finish: 62.5mm (total thickness) Kingspan Kooltherm K18, or similar, insulated plasterboard with skim finish. External finish: 18mm OSB sheathing board, Tyvek Housewrap, or similar, breather membrane, 50mm x 25mm timber batten, 12mm proprietary render board, render finish. The above proposed construction will provide a u-value of 0.3 W/m^2K. (Building Regulations AD Part L minimum requirement: u-value of upgraded wall construction not to exceed 0.30 W/m^2K.)	
3.3	**External Windows and Doors** UPVC clear double-glazed windows. Low-emissivity glass, with argon-filled double-glazed units. Elevation drawings to be provided for client's / architect's approval. Glazing to critical areas to be safety glass in accordance with current Building Regulations. Opening windows to open greater than 30 degrees (for ventilation). All new windows to habitable rooms to be fitted with trickle ventilators for background ventilation, capable of minimum 5000 sq mm equivalent area. Double glazing to achieve a 'U' value not exceeding 1.6W/m^2K (band C WER).	
3.4	**Lintels** All lintels to have minimum 150mm bearing at each end. Lintel specification TBC.	

16 North Drive
Page 3 of 8

CATRIONA LONGWILL, ARCHITECT
3 SOUTHGATES DRIVE, FAKENHAM, NORFOLK, NR21 8AQ
T: 01328 855 508 E: CAT@CATRIONALONGWILL.CO.UK
WWW.CATRIONALONGWILL.CO.UK

Figure 5.2 Example extract from a specification

DID YOU KNOW?

On one recent domestic project, waterproof plasterboard was specified for the new en-suite shower room but the approved plans and specifications were not checked and standard plasterboard was used instead. When the error was spotted, the plasterboard had to be taken down and replaced, wasting time, money and resources.

PRACTICAL TIP

Some plasterers wear latex medical gloves to protect their hands from the plaster. These are fairly inexpensive if bought in bulk online. They are not a replacement for thick safety gloves.

Personal protective equipment (PPE) for applying plaster materials

The risk assessment will identify how health and safety risks should be reduced to prevent the need for PPE. It may also suggest collective protective measures, such as scaffolds, which protect more than one person. Remember that PPE is always the last resort. However, it is usually still necessary to wear appropriate PPE.

You should always wear eye protection when cutting plasterboard. You are likely to create dust and fragments of the plasterboard may fly off. These are potentially very dangerous, particularly to the eyes.

Figure 5.3 Example of a collective protective measure to prevent plasterers falling from height

REED TIP

Look out for your mates. If they've forgotten to put on a piece of PPE, remind them!

You should always wear some kind of hand protection too. The thickness of the gloves will depend on the job, as you may need to be able to hold things, which can be difficult if your gloves are too thick. It may seem inconvenient and hinder your work but suitable thick gloves will protect your hands against cuts, abrasions and most impacts.

Using plaster and cutting plasterboard may require respiratory protective equipment (RPE) such as dust masks.

More information about suitable PPE and the laws concerning it can be found in Chapter 1 on pages 31–33.

Tools and equipment

You will need different tools for different jobs, from initial drawing and measuring, to attaching plasterboard or finishing the final coat of plaster. Whatever you are using them for, always use the right tool for the job and treat it with respect.

* Buy the best quality tools you can afford – cheap tools are a false economy as they are more likely to break or wear out, so you would need to replace them more often.

* Put tools away when you have finished with them.

* Always ensure you clean your tools thoroughly soon after you have used them. Plaster is difficult to remove once it has dried.

PRACTICAL TIP

It's worth getting a toolbag to keep all your tools together. Remember to keep an eye on your toolbag because if you lose it, you lose all your tools. Don't leave it in a van overnight as thefts from tradespeople's vans are common.

Drawing and measuring equipment

Measurement or marking tool	Description
Callipers and dividers 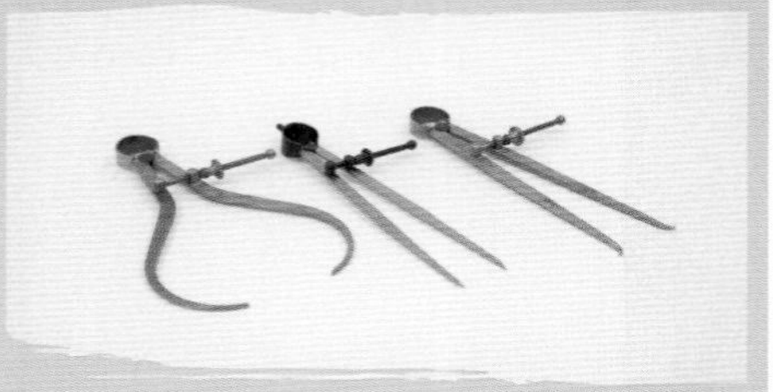 Figure 5.4	These are used to set out measurements.
Combination square  Figure 5.5	This is a useful tool that can do the job of a mitre square, tri-square and spirit level. It is useful for checking angles.
Eraser 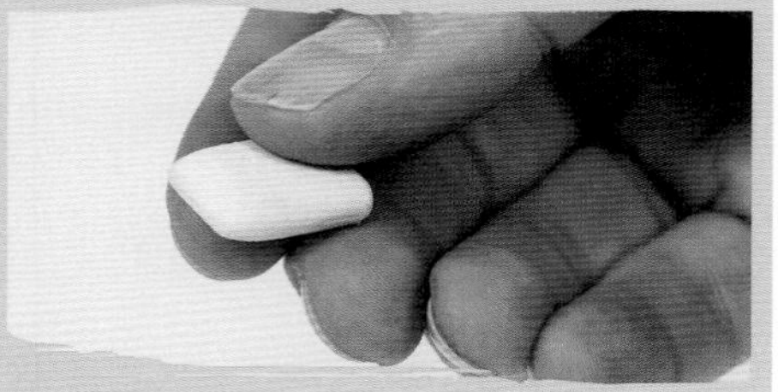 Figure 5.6	You will need an eraser to rub out any pencil marks that you have to change or that will show at the end of the job.
Levels 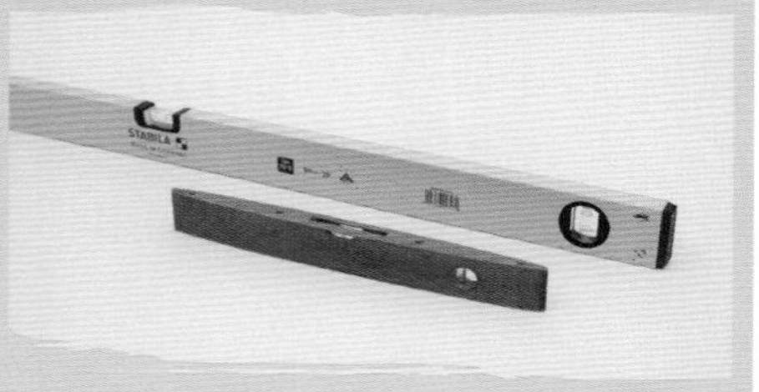 Figure 5.7	Levels are for checking the accuracy of work vertically and horizontally. Spirit levels contain a bubble of air in water – when the bubble is central, the work is level. They come in various sizes. Laser levels are increasingly popular as they are accurate and easy to use. It is useful to have both types of levels available.
Marking knife  Figure 5.8	This is considered to be better than a pencil as it is more accurate. It produces a slight cut that is useful when you get to the stage of cutting the plasterboard.

Measurement or marking tool	Description
Pencils Figure 5.9	These are graded according to the hardness or softness of the lead. You could use a carpenter's pencil but a normal pencil is fine as long as you keep it sharpened to provide a clean line.
Rules and measures Figure 5.10	Various options are available. Standard steel tape measures, folding rules and metal steel rules are all used. Although imperial measurement (inches) has been replaced by metric measurement (millimetres), it is wise to have rules and measures that show both. Make sure you don't get them mixed up.
Sliding bevel Figure 5.11	This is an adjustable tri-square. It is used to mark out angles other than those of 90°. You can set the blade at a particular angle and then lock it into place.
Tri-square or builder's square Figure 5.12	These are used to mark and test angles at 90°.

Table 5.1 Types of drawing and measuring tools used in plastering

Plastering hand tools

Tool	Description
Applicator 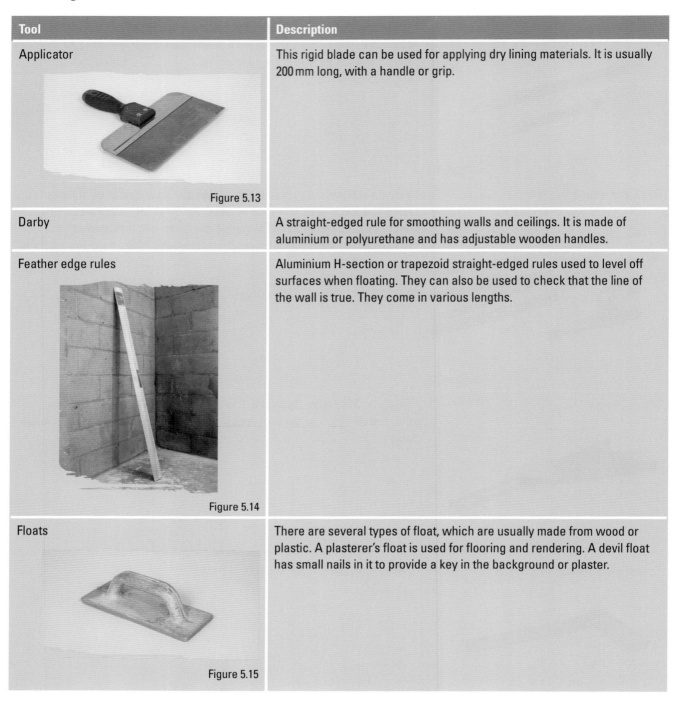 Figure 5.13	This rigid blade can be used for applying dry lining materials. It is usually 200 mm long, with a handle or grip.
Darby	A straight-edged rule for smoothing walls and ceilings. It is made of aluminium or polyurethane and has adjustable wooden handles.
Feather edge rules Figure 5.14	Aluminium H-section or trapezoid straight-edged rules used to level off surfaces when floating. They can also be used to check that the line of the wall is true. They come in various lengths.
Floats Figure 5.15	There are several types of float, which are usually made from wood or plastic. A plasterer's float is used for flooring and rendering. A devil float has small nails in it to provide a key in the background or plaster.

Tool	Description
Handboard/hawk 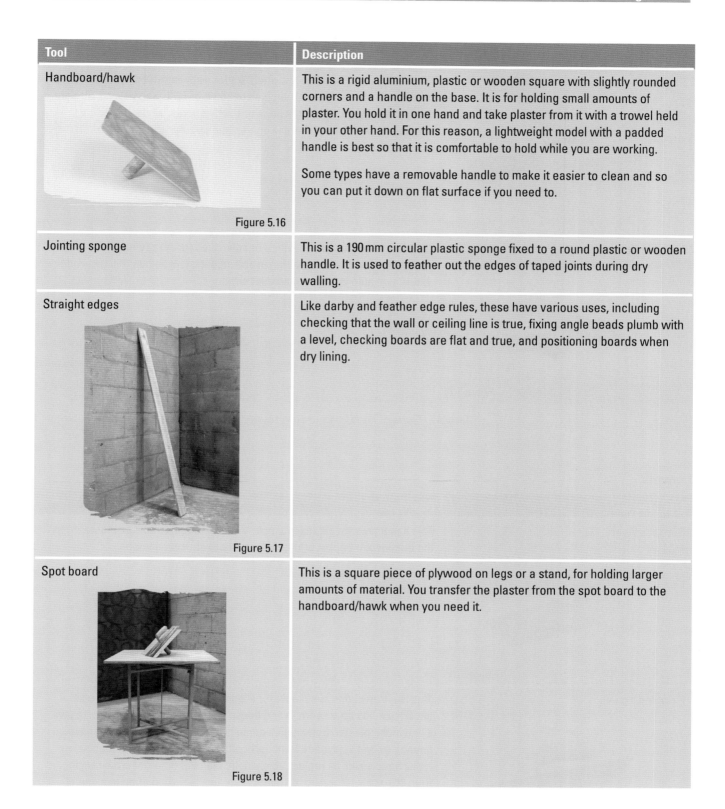 Figure 5.16	This is a rigid aluminium, plastic or wooden square with slightly rounded corners and a handle on the base. It is for holding small amounts of plaster. You hold it in one hand and take plaster from it with a trowel held in your other hand. For this reason, a lightweight model with a padded handle is best so that it is comfortable to hold while you are working. Some types have a removable handle to make it easier to clean and so you can put it down on flat surface if you need to.
Jointing sponge	This is a 190 mm circular plastic sponge fixed to a round plastic or wooden handle. It is used to feather out the edges of taped joints during dry walling.
Straight edges Figure 5.17	Like darby and feather edge rules, these have various uses, including checking that the wall or ceiling line is true, fixing angle beads plumb with a level, checking boards are flat and true, and positioning boards when dry lining.
Spot board Figure 5.18	This is a square piece of plywood on legs or a stand, for holding larger amounts of material. You transfer the plaster from the spot board to the handboard/hawk when you need it.

Tool	Description
Tin snips 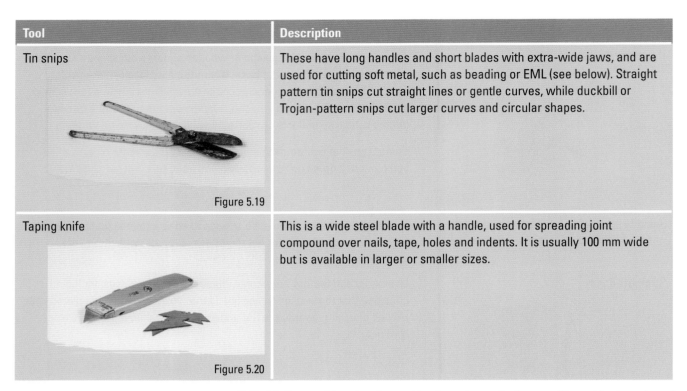 Figure 5.19	These have long handles and short blades with extra-wide jaws, and are used for cutting soft metal, such as beading or EML (see below). Straight pattern tin snips cut straight lines or gentle curves, while duckbill or Trojan-pattern snips cut larger curves and circular shapes.
Taping knife Figure 5.20	This is a wide steel blade with a handle, used for spreading joint compound over nails, tape, holes and indents. It is usually 100 mm wide but is available in larger or smaller sizes.

Table 5.2 Types of hand tools used in plastering and when working with plasterboard

Trowels

Type of trowel	Description
Bucket trowel 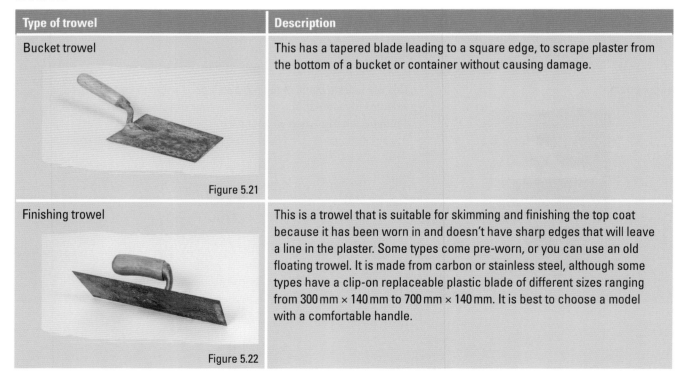 Figure 5.21	This has a tapered blade leading to a square edge, to scrape plaster from the bottom of a bucket or container without causing damage.
Finishing trowel Figure 5.22	This is a trowel that is suitable for skimming and finishing the top coat because it has been worn in and doesn't have sharp edges that will leave a line in the plaster. Some types come pre-worn, or you can use an old floating trowel. It is made from carbon or stainless steel, although some types have a clip-on replaceable plastic blade of different sizes ranging from 300 mm × 140 mm to 700 mm × 140 mm. It is best to choose a model with a comfortable handle.

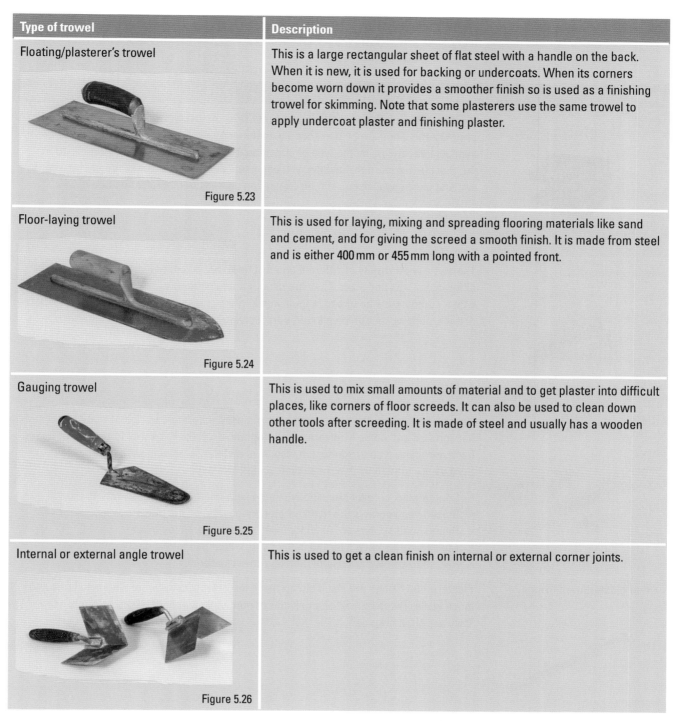

Type of trowel	Description
Floating/plasterer's trowel Figure 5.23	This is a large rectangular sheet of flat steel with a handle on the back. When it is new, it is used for backing or undercoats. When its corners become worn down it provides a smoother finish so is used as a finishing trowel for skimming. Note that some plasterers use the same trowel to apply undercoat plaster and finishing plaster.
Floor-laying trowel Figure 5.24	This is used for laying, mixing and spreading flooring materials like sand and cement, and for giving the screed a smooth finish. It is made from steel and is either 400 mm or 455 mm long with a pointed front.
Gauging trowel Figure 5.25	This is used to mix small amounts of material and to get plaster into difficult places, like corners of floor screeds. It can also be used to clean down other tools after screeding. It is made of steel and usually has a wooden handle.
Internal or external angle trowel Figure 5.26	This is used to get a clean finish on internal or external corner joints.

Table 5.3 Types of trowels used in plastering

You will also need brushes and buckets like those described in Chapter 4.

Mechanical fixing aids

These are the fixings you use to attach plasterboard to the background surface.

Fixing	Description
Plugs  Figure 5.27	Plastic wall plugs are screw fixing devices. They are usually made either from nylon or polythene. They are colour-coded to match screw gauge sizes. Different types are available depending on the type of plasterboard and background. For example, to fix thermal plasterboard to masonry, you can use a plug that is a combination of masonry nail and plastic wall fixing with an expanding tip and countersunk head.
Nails  Figure 5.28	There is a wide variety of different types of nails, not only for different jobs but also in different lengths and sizes. They are usually made of non-corrosive galvanised steel, with a countersunk head and jagged shank to provide a secure fix. Due to the likelihood of nails 'popping' out of the plasterboard, screws are generally seen as a better option.
Screws  Figure 5.29	Drywall screws are the most common fixing used for attaching plasterboard to a timber background and come in several sizes, depending on the thickness of the plasterboard. Each particular type of screw can also have a variety of driving slots or recesses. Screws are graded by their head type, length and gauge: • Countersunk screws are used when the screw needs to be flush with the work. • Raised head screws are used for attaching metal components. • Round head screws are used for fixing sheet material to timber. • Mirror screws have a thread in the head, which can take a decorative dome. • Pan head screws are useful for fixing sheet material.
Other plasterboard fixings 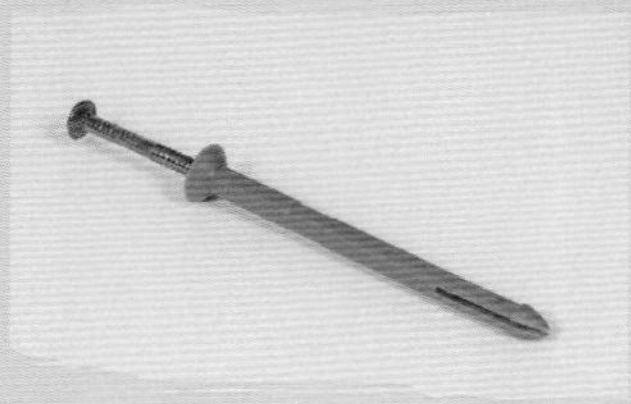 Figure 5.30	Plasterboard fixings are available for fixing insulated plasterboard sheets to brick and block walls. These consist of a galvanised mechanical anchor with a concave head. They are often fireproof.
Adhesives  Figure 5.31	Plasterboard adhesive (also called bonding compound) is usually a gypsum-based powder that is mixed with clean water to create a strong enough bond for attaching plasterboard to brick, block, sand and cement backgrounds. Standard plasterboard adhesive is not waterproof and is not strong enough to be used on ceilings. It has a typical working time of about 100 minutes.

Table 5.4 Types of mechanical fixings used in plastering and plasterboarding

Portable powered tools

Portable power tool	Description and use
Jig saw  Figure 5.32	Although you would normally use a trimming knife or basic handsaw, a jig saw can be useful for curved work and odd shapes. A number of different blades are available.
Drill/electric screwdriver  Figure 5.33  Figure 5.34	These are battery or mains operated. They usually have two different settings and speeds, depending on what you are using the drill for. Some have a hammer action that strikes into solid backgrounds like concrete. Use the speed that is most appropriate to what you are doing. Slow speeds make a more accurate hole and prevent damage to begin with, before moving up to a faster speed.
Drywall sander Figure 5.35	This tool has large diameter discs for sanding drywall joints, painted wall surfaces and ceilings. Hand-held sanders (like the one here) are used for small areas. Each type of sander leaves a more or less smooth surface, so it is important to use the right belt or paper for the job and the material being sanded.

Portable power tool	Description and use
Screw gun 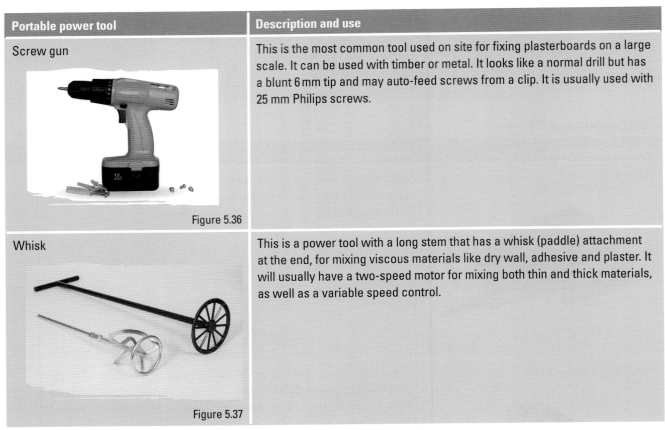 Figure 5.36	This is the most common tool used on site for fixing plasterboards on a large scale. It can be used with timber or metal. It looks like a normal drill but has a blunt 6 mm tip and may auto-feed screws from a clip. It is usually used with 25 mm Philips screws.
Whisk Figure 5.37	This is a power tool with a long stem that has a whisk (paddle) attachment at the end, for mixing viscous materials like dry wall, adhesive and plaster. It will usually have a two-speed motor for mixing both thin and thick materials, as well as a variable speed control.

Table 5.5 Types of power tools used in plastering and plasterboarding

Ancillary equipment

As a plasterer, you will often be working alone, so you cannot depend on someone else being able to help you reach high areas or fix heavy plasterboard. Trying to lift or hold plasterboard yourself can lead to back injuries. Equipment is available to take away the strain. However, always make sure you know how to use it in advance and don't take unnecessary risks.

Equipment	Description
Board trolley/drywall trolley	These robust steel-frame trolleys are designed for easy handling of sheet and board materials. They are narrow enough to easily go through a doorway and are ideal for areas with uneven or soft flooring.

Equipment	Description
Foot lifter  Figure 5.38	This slides under the plasterboard so that you can lever it off the floor to a height of 60 mm or so.
Panel lifter/board lifter	This is a mechanical device to enable one person to lift plasterboard and panels into position onto ceiling and wall studwork. Larger ones can support 75 kg. The board can be raised up to 4 m, swivelled through 360° and tilted to 45°. Often they have a solid frame that can be dismantled for transporting.
Panel prop/dead man/panel jack	This telescopic prop or jack holds plasterboard in position, freeing your hands to fix it without having to take its weight yourself. The equipment's adjustable height enables you to use it in different types of room. It can normally support up to 45 kg. Some models have a sprung quick release trigger and head that springs the top plate into position after triggering the release button.
Step up/hop-up  Figure 5.39	These are lightweight aluminium platforms for short-duration work on walls and ceilings. They are normally about 80 cm high and about 150 cm long. Larger ones can support up to 150 kg. Models with handrails are safer.
Stilts  Figure 5.40	These give you some extra height to reach upper walls and ceilings. They are made to fit people of all sizes and have adjustable heights, calf straps and heel plates. Common heights are 450 mm to 760 mm and 610 mm to 1,000 mm.

Table 5.6 Ancillary equipment

Types of plaster

Gypsum plasters

Plaster must comply with British Standard 1191, the specification for **gypsum** building plasters. Some plasters are grey and others are pink, depending on where they come from, but the colours do not matter as long as they comply with BS 1191.

Most gypsum-based plasters now use lightweight aggregates like exfoliated **vermiculite** and expanded perlite instead of sand. The aggregates are mixed with Class B gypsum plasters for undercoats and finish coats. They can be used anywhere that sand-based plasters would be used, and in fact have better heat resistance and thermal insulation qualities than sand-based plasters so can be used where condensation may be a problem. They come pre-mixed for different applications, and only need water to be added.

Lightweight plasters are:

* 60 per cent lighter than sand plasters

* 300 per cent more insulating than sand plasters

* fire resistant

* crack resistant

* able to stick to smooth concrete and to glazed or oil-painted surfaces.

Choosing your plaster

In new work, the type of plaster you use depends on:

* the background materials

* the suction of the background

* the required hardness of the finished plaster.

Most plasters are produced by British Gypsum and Knauf, and have recognised brand names that you will soon become familiar with. British Gypsum provides the majority of material for the plastering industry in the UK, but new technology and easier production has meant that other companies are competing to produce materials for plaster. These are now being used around the country.

Table 5.7 describes some of the plasters that you may use. Even if you don't use British Gypsum plasters, it is useful to know the variety of properties that different plasters can have.

Type of plaster	Name	Purpose
Undercoats	Thistle Bonding Coat	For smooth or low suction backgrounds, e.g. concrete, plasterboard or surfaces pre-treated with bonding agents. You can also use it to fix EML.
	Thistle Hardwall	Has high-impact resistance and a quicker drying surface. Suitable for application by hand or a mechanical plastering machine to most masonry backgrounds.
	Thistle Tough Coat	Has high coverage and good impact resistance. Suitable for application by hand or a mechanical plastering machine to most masonry backgrounds.
	Thistle Browning	For solid backgrounds of moderate suction with an adequate mechanical key.
	Thistle Dry-Coat	Is cement based, for re-plastering after the installation of a damp-proof course.
Finish coat plasters	Thistle Board Finish	For skimming low–medium suction backgrounds such as plasterboard.
	Thistle Multi-Finish	For use over both undercoats and plasterboard.
	Thistle Uni-Finish	A finish coat plaster that requires no prior preparation with PVA on the majority of backgrounds.
	Thistle Spray Finish	A gypsum finishing plaster for spray or hand application.
One coat plaster	Thistle Universal One Coat	For a variety of backgrounds. Suitable for application by hand or a mechanical plastering machine.

Table 5.7 Types of British Gypsum plasters. *Source: The White Book*, 2013

PRACTICAL TIP

Brand names can change so always check the bag before you mix the plaster to make sure you are using the right product.

DID YOU KNOW?

Plaster is usually applied by hand but a spray application is often preferred when a large area needs to be covered on commercial projects, such as on a long corridor.

Figure 5.41 Transferring plaster from the handboard to the trowel

Non-gypsum plasters

Damp walls, or walls that have to be re-plastered after a DPC has been installed, are not suitable backgrounds for gypsum plasters as they will draw salts to the surface of the plaster. Instead, plasters containing well graded sand and cement should be used. However, these are brittle and dense so may crack with building movement, so breathable **lime**-based plasters are another alternative. Refurbishment work is likely to use lime plaster, as this is usually the type found in older buildings. These are a 1:1:6 cement/lime/sand plaster mix with a lightweight aggregate used in place of sand.

Ensuring compatibility between backgrounds and applied plaster

As we saw in Chapter 4, the type of background determines its suction and you need to control this while preparing to plaster. You must also make sure you use the correct type of plaster for the job – see Table 5.7 above for descriptions of different types of plasters.

The specification should also tell you what types of plastering materials to use, along with the required thickness and number of coats. These are also identified on bags of plaster.

Starting to work with tools and materials

Before you can achieve the skills to produce a wall plumb and straight you must first learn the art of working with the materials. Tool manipulation is the first skill to master. Here are some basic points to remember.

* Always place the mixed material on a board. **Do not use materials straight from a bucket** (except for small patch work).

* Ensure the material is a good consistency.

* Keep tools clean.

* Practise taking the plaster from the handboard to the trowel.

* It is best to use lime sand mortar at first, as it takes longer to set so will give you time to adjust to the workability of the material.

* Watch experienced tradespeople, and listen and learn.

Applying one, two and three coats of plaster

Internal plastering may consist of one, two or three coats, depending on the background surface and the type of work required. The plaster is normally applied in two coats – the first coat (or undercoat) is about 11 mm thick, with a finish coat (or skimming coat) over it that is about 2 mm thick – but the background or function of the plaster may require one or three coats to be applied.

One coat work

All-in-one plasters are sometimes used, and are applied thickly, often using a spray applicator, without a second, finish coat. They may be suitable in old buildings where a completely smooth wall is not required and for patching small areas. However, for standard brickwork walls, most professional plasterers feel that this type of plaster and approach doesn't give an adequate finish.

You are most likely to use a single coat to provide a thin **finish**, skim or setting coat to plasterboard or other flat sheet materials. It is usually 2 mm or 3 mm deep and is usually applied in stages. Finish coat plasters are finer than undercoat plasters and can create a very smooth surface – if they are applied with skill. Applying a skim coat is described under 'Two coat work' below, and the procedure is the same as skimming onto a floating coat.

Two coat work

This (also known as float and set) is most plasterers' preferred method of internal plastering a solid background like brick or blockwork. It involves applying an undercoat or **floating coat** for a flat surface, which is keyed before a finish (skim or setting coat) is applied to give a smooth surface.

Apply the floating coat to flat surfaces up to a thickness of 5 mm or by hand to 13 mm.

Unlike with one coat plastering, you will be applying your second (finish) coat to plaster (a floated background) rather than plasterboard. This means that, before applying the finish coat:

* you must make sure that the floating coat is not yet completely dry

* you must ensure you have keyed the floating coat, preferably with a devil float

* you may need to damp down the background with water if suction is too low

* you may need to scrape down the background, especially internal and external corners, which may need cutting back a few millimetres from the angle, using the back of a trowel.

See below for methods of applying the floating coat – plumb, dot and screed, and broad screed.

KEY TERMS

Finish coat

– finish plaster applied 2–3 mm thick to provide a smooth finish ready for decoration.

Floating coat

– undercoat plaster, commonly lightweight, applied 8–11 mm thick to a background to make it straight and plumb prior to the setting coat being applied.

PRACTICAL TIP

A common plaster for two coat plastering is Thistle Multi-Finish. This is a gypsum finish plaster that can be used on a range of backgrounds. It is popular because it is easy to prepare and apply, and provides a smooth, attractive surface to internal walls and ceilings.

Applying the finish coat

Apply the finish coat (also called the finishing, skim, skimming or setting coat) with a finishing trowel, starting from the top left side if you are right-handed, and from the right if you are left-handed.

When you apply the plaster, ensure you overlap each trowelful, so that you do not leave any gaps, then go over it to leave a coat about 1mm thick initially. Then you trowel up by wetting a clean trowel and firmly and methodically smooth the wall, holding the trowel at about 45°. The final coat should be no thicker than 5mm.

PRACTICAL TIP

When you have a grainy surface like sand or cement, or plasterboard that needs flattening, you can apply an additional thick layer using the float to cover up any irregularities. You can also rule it off with a feather edge rule.

You should aim to leave a matt finish, as a shiny finish won't provide enough of a key for the subsequent paint. You can do this by brushing clean water on the wall in front of your trowel as you finish the skim coat.

When you have finished, clean out all the excess plaster from internal and external angles, the ceiling line, electrical points and the skirting line.

You will often see that plaster is not applied below the skirting line. This is to enable the skirting to be fixed flush to the wall. Mark off a line about 25mm from the floor and remove the plaster below this line.

Figure 5.42 A skim coat

Figure 5.43 Applying two coat plaster

Figure 5.44 Multi-finish plaster

Three coat work

Sometimes the background requires an extra coat of plaster material. This might be because:

* it is uneven

* it contains metal lathing such as EML, beads and trims

* it is made of several different materials

* the area is severely damaged, for example around walls and windows in older properties

* the plaster needs to be thicker, as specified by the architect or because you need to match an existing thickness.

Three coats consist of two undercoats and a finishing coat, in a process called render, float and set:

1. **Scratch coat** or render undercoat to provide a solid background and adequate suction. It is keyed and left to dry for 24 hours.

2. Floating coat

3. Finish coat.

Three coat work should be applied as layers of up to 15 mm, which are left to dry before the next layer is applied. Once it has set (usually after a few days), you can add the key or scratch coat.

You can apply your floating coat using the plumb and dot method (also known as the plumb, dot and screed method) or the broad (or box) screed method. Plumb and dot is more complicated but is thought to produce a better result than the quicker broad screed method.

KEY TERMS

Scratch coat

– the first coat of plaster materials applied, to control suction, straighten and even out walls and provide a mechanical key for the next coat.

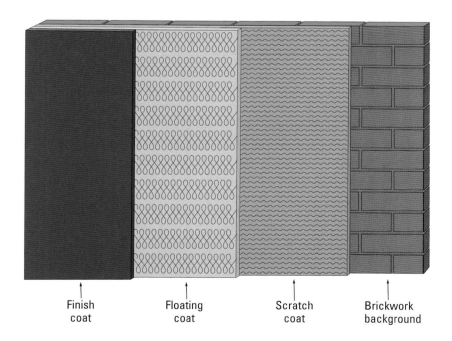

Finish   Floating   Scratch   Brickwork
coat   coat   coat   background

Figure 5.45 Example of three coat plastering

KEY TERMS

Dots

– small pieces of material such as plywood applied to the background to help produce a flat floated surface. They form guidelines for the floating rule when applying screeds.

Screeds

– narrow bands of plaster material. They are built up between plumbed or levelled dots, to act as guides for the floating rule when creating plumb or level surfaces on walls and ceilings.

PRACTICAL TIP

If your dots are made from timber, soak them in water before use.

PRACTICAL TIP

The dots will usually be 12 mm thick when bedded to the wall.

Using the plumb, dot and screed method

The plumb, **dot** and **screed** method produces the most accurate result but it is fairly time-consuming.

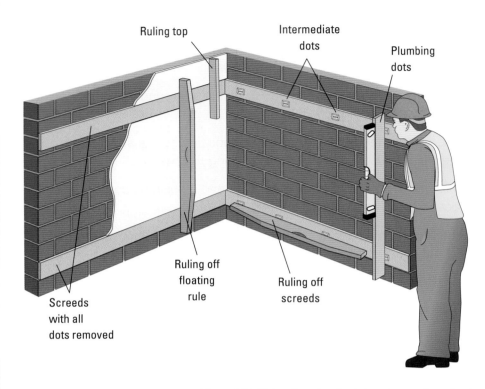

Figure 5.46 Using the plumb, dot and screed method

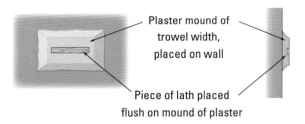

Figure 5.47 A dot

Before you start, check whether there is a door or opening in the wall. If so, you will need to set your dots to the same level to ensure the wall doesn't go too far back or forward.

Set a dot approximately 150 mm from a bottom corner of the wall, bedding it in with a little plaster. Then set a dot 150 mm from the other bottom corner. It's worth setting a line between them to check whether they are level. If not, adjust the dots.

Once they have set, apply dots in the top corners. Use a level to check they are plumb with the bottom dots.

Now set more dots in a row above the bottom dots, using a string line to ensure they are straight. These should be about 1,500 mm apart but ensure your straight edge can reach from one dot to the next. Repeat for the dots below the top row. You should now have four rows of dots.

After they have set, form the screeds by applying a band of plaster vertically between two of the dots, making sure it sticks out further than the dots. With your rule, push the plaster back so that it is flush with the dots and use your trowel to cut off the surplus plaster on each side of the rule. Carefully slide your rule off the screed and continue until you have applied screed between all the dots.

Once the screed has set, you need to fill in the bays (the sections between the screeds). Damp the screeds down to control the suction then float one bay at a time, using your rule as before and cutting off the excess plaster. Fill in any hollows or gaps and rule it off again before going on to the next bay.

Keep checking your wall to make sure it is flat and plumb.

When you have finished every section and it has set, key the wall using a devil float.

Floating to wet screed using the broad screed method

A problem with the plumb, dot and screed method is that you have to wait for the plaster to set, which can cause problems on site when there is a tight schedule. This means you often need to apply your floating coat quickly. The broad screed method of floating and ruling it off is preferred by many plasterers on site for this reason, but the finish will not be to as high a standard.

1. Apply the plaster to a corner section of wall and pass the rule across it until the plaster is consistently around 11 mm thick.

2. Pass the rule over the plaster using an up-and-down movement until it is flat and smooth. If you have missed any areas, fill them in and re-rule them.

3. Trim off any excess plaster from ceilings, angles and wall lines using a trowel, and clean these areas with a brush.

4. Repeat for the opposite corner of the wall.

5. Apply the floating coat from right to left if you are right-handed, or left to right if you are left-handed. Rule off the wet screeds using a straight edge. Keep cleaning off the rule to make sure your edge remains sharp.

6. When you have ruled in the wall flat, check it is flat and plumb with a level and correct any irregularities.

7. Rub up with devil float and clean all angles with a trowel.

PRACTICAL TIP

Check for straightness and plumb to 3 mm within each 1.8 m length.

DID YOU KNOW?

You can make your own devil float by pushing four or five screws into a polystyrene plastering float, so that the points of the screws protrude by about 2 mm.

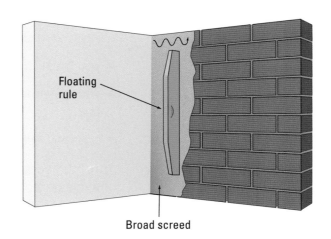

Floating rule

Broad screed

Figure 5.48 Using the broad screed method

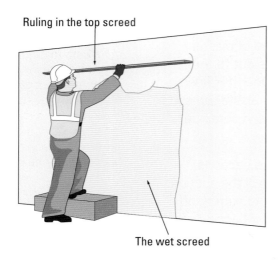

Ruling in the top screed

The wet screed

Filling in the bottom screed

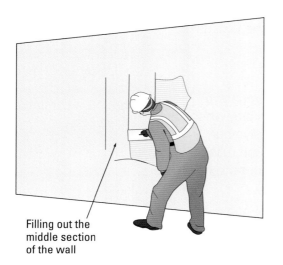

Filling out the middle section of the wall

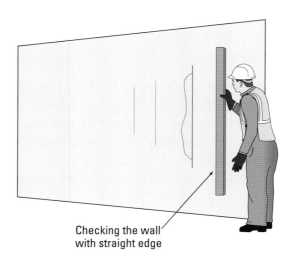

Checking the wall with straight edge

Figure 5.49 Applying a floating coat using the broad screed method

Applying two coat plaster

Unlike with one coat plastering, you will be applying your second (finish) coat to plaster (a floated background) rather than to plasterboard. This means that, before applying the finish coat:

* you must make sure that the floating coat is completely dry

* you may need to damp down the background with water if suction is too low

* you may need to scrape down the background.

Apply the second coat in the same way as you would when skimming onto plasterboard.

Always leave the area clean and tidy. For example, put down a protective sheet to prevent plaster from getting onto the floor. After you have finished, clean out electrical sockets and ensure all doors and window frames and glazing are free from plaster.

PRACTICAL TIP

Try to avoid working to wet angles by completing the opposite walls first. When these two walls have set the other two opposite walls can be completed.

Working with plasterboard

Plasterboard has several advantages over plaster.

* It provides better insulation than plaster. This is because of the layered materials it is made from, and because dry lining creates an air gap between the plasterboard and the background surface, which traps heat.

* It is quicker to build a wall from plasterboard than from bricks or blocks.

* You don't need to wait for plasterboard to dry before applying a finish. For this reason, it is often used as an alternative to two coat plastering work.

All these are economic advantages because the project is finished sooner, and energy bills are lower for the occupier.

> **REED TIP**
>
> Working as a plasterer is about people, not just technical skills. You won't just be making repairs, for instance, you'll be delivering a service to a customer.

> **PRACTICAL TIP**
>
> Do not use plasterboard to cover up a damp wall. The problem has not been solved, and the wall could become worse, with the damp showing through the board.

Dry lining

The terms plasterboarding and dry lining are often used interchangeably. Dry lining gets its name because a dry material (plasterboard) is applied to an existing wall, rather than a wet material (plaster). Confusingly, sometimes dry lining refers to the **direct bond** application method, which fixes the plasterboard to the background using wet plaster and adhesives rather than mechanical fixings.

> **KEY TERMS**
>
> **Direct bond**
>
> – the fixing of plasterboards or beads with dabs of plaster or adhesive.

Types and sizes of sheet material

The properties of plasterboard vary according to the additives in the gypsum layer and the weight and strength of the lining paper. You need to make sure you use the correct type of plasterboard to match the specification. As well as standard plasterboard, which complies with fire and thermal standards and regulations, plasterboard with other qualities is available.

> **PRACTICAL TIP**
>
> Applying plasterboard to a timber frame to form a new wall is not dry lining because it is not attached to an existing background surface.

> **DID YOU KNOW?**
>
> Fibre-reinforced gypsum boards (like MultiBoard) are not strictly classed as plasterboard because they don't have a paper facing. However, they are still a plaster sheet material that you may be required to fix.

Figure 5.50 Plasterboard showing the plaster between the paper faces

Most standard plasterboard has one ivory face, which is suitable for having plaster applied to it, and one brown (reverse) face. The properties of each specialist board are indicated by having a different colour on its face, making it easy to tell what sort of board it is.

Sizes

Plasterboard comes in many different shapes, sizes and thicknesses. It usually comes 1,200 mm wide, to suit the standard 600 mm stud spacing used in modern housing, but sheets can be as narrow as 600 mm for use where space is limited. Of course, you can also cut it to size. The thickness depends on its purpose – for example, standard plasterboard is available in a 9.5 mm thickness, but insulating plasterboard can be ten times thicker than that.

Brands

There are many different brands of plasterboard and each plasterer has their favourite. The brands listed in Table 5.8 are produced by British Gypsum (Gyproc) but you will also see plasterboard made by companies like Knauf and Lafarge. If a brand is specified by the architect, you must use it. If not, choose the type that best matches the specification.

Type of board	Example brand name	Colour of facing paper	Uses	Available sizes
Standard performance	WallBoard or HandiBoard	Ivory	This is suitable for most applications where minimum fire, structural and acoustic levels are specified. It often enables plaster or decoration to be applied directly.	Thickness: 9.5 mm, 12.5 or 15 mm Size: 600 mm × 1,220 mm to 900 × 1,800 mm
Baseboard	Baseboard	Grey (no facing paper)	This is a cheap plasterboard with no ivory facing or insulation. It can be attached to the wall or ceiling to create a flat surface either for a specialist plasterboard or for direct plastering. It can also be used as a trim at the base of a wall, below the main plasterboard.	Thickness: 9.5 mm
Fire-resistant	FireLine	Pink	This is a non-combustible glass-reinforced gypsum board, which gives increased fire protection.	Thickness: 12.5 mm to 30 mm Size: 900 mm × 1,800 mm to 1,200 mm × 3,000 mm
Glass-fibre reinforced	MultiBoard	Varies	In this board, glass fibre is used between the layers instead of paper. It is suitable for constructing all forms of partitions and ceilings, including curves, with fire and impact protection.	Thickness: 6 mm, 10 mm or 12.5 mm Size: 1,200 mm × 2,400 mm to 1,200 mm × 3,000 mm

Type of board	Example brand name	Colour of facing paper	Uses	Available sizes
Insulating	ThermaLine	Varies	This board retains heat and controls vapour.	Thickness: 38 mm to 93 mm Size: 1,200 mm × 2,400 mm
Moisture-resistant	Tilebacker	Green Ivory for vapour check boards	This board is used in damp areas like wet rooms and showers, and as a base for ceramic tiles.	Thickness: 6 mm or 12.5 mm Size: 1,200 mm × 900 mm to 1,200 mm × 3,000 mm

Table 5.8 Types of plasterboard

PRACTICAL TIP

The terms 'wallboard' and 'baseboard' are sometimes used instead of 'plasterboard' so make sure you are sure you are using the right type before you start work.

Methods of fixing plasterboard

Fixing plasterboard using mechanical fixings

Mechanical fixings are used to secure plasterboard to walls and ceilings. You would not normally drill directly into the wall; the boards are attached to timber battens, joists or a stud wall frame.

Remember to stagger the plasterboard so that you don't have a long line of joints, which could weaken the wall or ceiling. The boards should be placed lengthways to lie horizontally across joints. Leave a gap of up to 5 mm between each board.

Figure 5.51 Standard plasterboard – ivory facing

Figure 5.52 Fire-resistant plasterboard – pink facing

Figure 5.53 Moisture-resistant and acoustic plasterboard – green facing

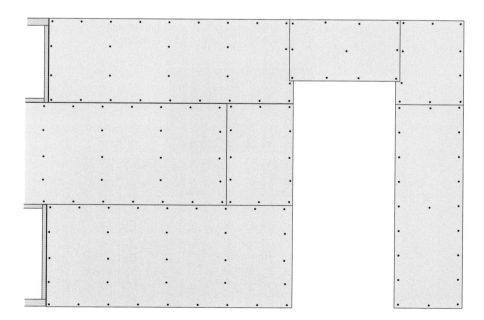

Figure 5.54 Staggering joints across the wall

Each board has two bound edges and two unbound edges. Make sure the long bound edge spans the joists at right angles and that the unbound edge is halfway on the furthest joist. This should keep the boards square.

Start fixing from an internal corner and work across to the external angles.

The size of the screws or nails you need depends on the thickness of the plasterboard. They need to attach firmly to the wooden or metal frame behind. Table 5.9 shows the relationship between the thickness of the board and the length of the screw.

Thickness of plasterboard	Drywall screws	Plasterboard clout nails
9.5 mm	32 mm	30 mm
12.5 mm	38 mm	40 mm
15 mm	38 mm	40 mm
19 mm	41 mm	50 mm
12.5 mm double boarded	51 mm	50 mm

Table 5.9 Size of fixings to be used with different thicknesses of plasterboard

Fix screws at 300 mm intervals (centres) and nails at 150 mm intervals, and position them more than 13 mm from the edge of the board.

Don't drive in the head of the nail too much or too little as this could risk puncturing the paper and damaging the plaster inside, so that the board is too loose or falls off. The nail should go in just enough that it grips the paper.

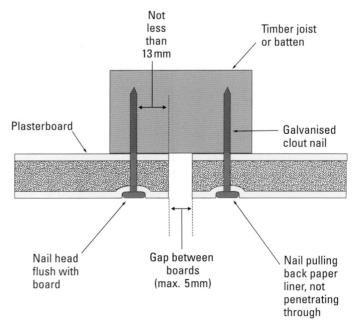

Figure 5.55 Correct positioning of nails

Fixing plasterboard using adhesives

This is known as direct bond, or the dot and dab method.

Direct bond has two main advantages over mechanical bonding:

* It can be bonded directly to bare bricks, instead of having to fix it to wooden battens.

* You don't need to make good the screw holes by covering them over.

These advantages speed up the job without compromising on quality.

You need to have a fairly flat wall to use direct bond – if not, consider using mechanical fixings attached to timber battens.

The type of adhesive you use for the job depends on the type of application and the desired completion time. You will normally use powdered pre-mixed drywall adhesive. This is suitable for all types of background and suction, though it is still good practice to paint the wall with PVA or another primer first. Following the instructions on the package, add the adhesive slowly to water and mix it in a well-ventilated area with a heavy duty drill or whisk. Leave it to stand for a few minutes and give it a stir before using it.

Reinforcing joints on plasterboard

To reduce the risk of cracks on plasterboard joints, corners, walls and ceilings, you may need to apply scrim or tape before you plaster. The two terms are often used interchangeably but, generally, scrim is used when the wall is going to be plastered and tape is used when the wall is going to be dry lined or where background plasterboards need to be joined together before plastering.

Both scrim and tape also seal the lining, which is vital in order to achieve specified levels of fire resistance and sound insulation.

Scrim tape is specified on some jobs. It is a flexible strip of mesh or cloth that is sticky on one side. Hessian or jute scrim gives extra strength to wall and ceiling angles. It usually comes in rolls of 50 mm × 90 m.

Figure 5.56 Scrim tape

Jointing tape is a strip of perforated paper with a sticky back. It comes in rolls of 50 mm × 90 or 150 m.

Other types of tape include:

* **waterproofing tape**: to stop water getting into joints in shower and bathroom areas

* **corner bead tape**: flexible paper tape with a metal strip along the centre to help create sharp edges when plastering

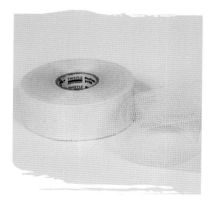

Figure 5.57 Jointing tape

* **wall repair tape**: ultra-thin self-adhesive fabric tape that can be painted over

* **rough surface joint tape**: double-sided tape for use on rough or bumpy surfaces.

> **DID YOU KNOW?**
>
> Dispensers are cheap to buy and make it easier to use scrim and tape. Some dispensers are designed to hang from your belt. You can also buy applicators that apply the tape directly to the background but these are much more expensive.

Using metal components

Metal trims like EML, rib lath and beading are used to make it easier to get a good finish after plastering.

EML

EML (expanded metal lathing) is a sheet of diamond-shaped mesh steel. It is used internally or externally to:

* provide a key on a variety of backgrounds for materials such as plaster, suspended ceilings and timber-framed buildings

* cover irregular or bumpy surfaces

* cover surfaces where two different materials meet (such as when plasterboard is next to the wood of a new door frame)

* reinforce corners

* make curved and free-form structures

* provide a carrier for fireproof finishes on structural steelwork.

Its flexibility and weatherproof properties give EML an advantage over plasterboard in projects where this is important. It is made from either galvanised or stainless steel and is supplied in sheets of 2,700 mm × 700 mm, although other weights and mesh sizes are available. It can be cut into the size you require. It is fixed to the wall using specialist stainless steel nails.

> **PRACTICAL TIP**
>
> When cutting EML, use tin snips or a small hacksaw and ensure you wear protective gloves and goggles.

> **KEY TERMS**
>
> **EML (expanded metal lathing)**
>
> – metal reinforcement made out of sheet metal to form a mesh, and fixed over concrete, timber or friable backgrounds to provide a key.

Beading

Beading or trims are used internally and externally as edgings to get a sharp corner, for example where the wall meets the ceiling. They can be used with plaster or plasterboard. Galvanised steel or uPVC are used for internal work. Stainless steel beading is used where there is a high moisture content, such as in bathrooms and kitchens, or outside the building.

Beading strips come in different sizes and are cut to size with tin snips or a hacksaw, or nailed together. They can then be bedded into the scratch coat.

Figure 5.58 Beading being bedded around a window opening

Beading has a number of benefits:

* It strengthens corners and edges, so that the plaster is less likely to be chipped.

* Ready-made edges mean that features like **arrises**, **stops** and **movement joints** don't need to be made by hand.

* They are an easier way of providing decorative features than trying to mould them yourself out of plaster.

DID YOU KNOW?

uPVC beading comes in white as standard but can be supplied in other colours like grey, brown, ivory or black.

KEY TERMS

Beading

– steel or plastic strips applied to create sharp edges, angles and openings.

KEY TERMS

Arris

– a sharp corner.

Stop

– a break between two different types of material.

Movement joint

– a break between sections of a structure or wall to allow for movement of the construction materials. This may be due to changes in temperature, vibration or ground movement. Also known as an expansion joint.

Table 5.10 describes various types of beading you may need to use for different purposes.

Type of bead	Use
Angle bead 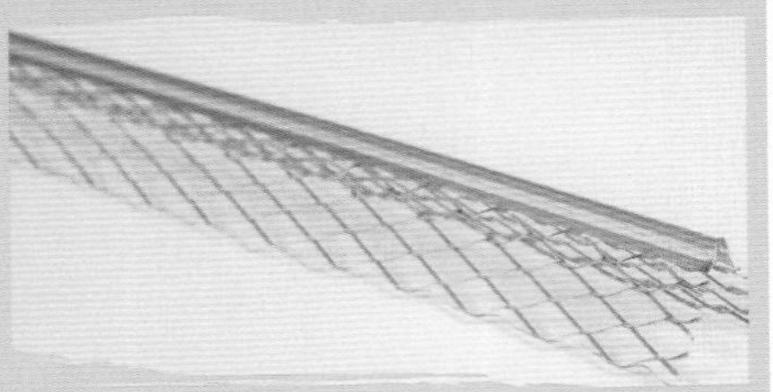 Figure 5.59	To form external angles and prevent chipping or cracking. It comes in different sizes and angles.
Architrave/shadow line bead	To form a clean division between different wall finishes, for example around door frames.
Movement bead  Figure 5.60	To allow for movement between adjoining surface finishes or sections that might move. It usually allows + or – 3 mm of movement. It can also be used where changes in the render colour are specified.
Stop bead  Figure 5.61	To provide neat edges to two coat plaster or render work at openings or abutments onto other wall or ceiling surfaces. Render (external) versions help with water run-off.
Thin coat bead 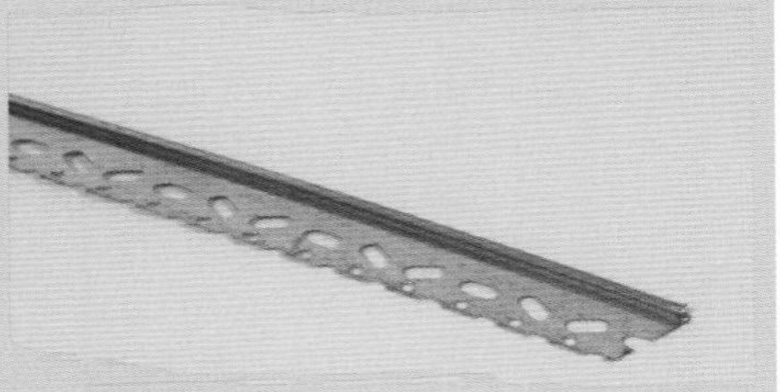 Figure 5.62	To be used with thinner coats of plaster and lightweight plasters. Angle and stop thin coat beads are available.

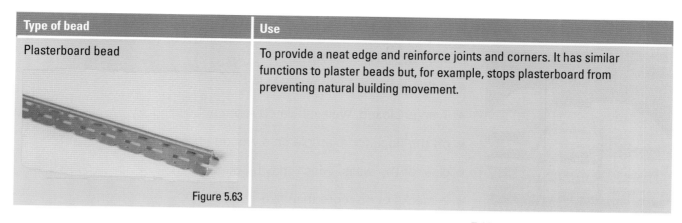

Type of bead	Use
Plasterboard bead	To provide a neat edge and reinforce joints and corners. It has similar functions to plaster beads but, for example, stops plasterboard from preventing natural building movement.

Figure 5.63

Table 5.10 Types of bead and their uses

PRACTICAL TIP

You can fix beading or mesh by securing it in place with mechanical fixings like steel nails, screws or staples then applying dabs of finish plaster. Wipe off any surplus plaster before it dries. If you don't bed it in at this stage, the adhesion may be reduced because it is difficult to squeeze plaster between the bead or mesh and the background surface.

Protecting the work and its surrounding area from damage

If you have to leave the work area before beginning to plaster or apply plasterboard:

* prevent people from accessing or interfering with your preparation work by closing off the area, if possible

* ensure that other trades are not scheduled to work in the area while you are away

* ensure the area is watertight – that windows and doors are closed and that it is free from leaks

* put away your tools and materials

* leave the area clean and tidy, disposing of waste according to the site rules or customer preference.

Storing plaster products

Plaster, like food, goes off if it is not used within its shelf life – before the date stamped on its bag. This shelf life should be three to four months if stored correctly.

Figure 5.64 Correctly stored bags of plaster

It will also be unusable if it is stored badly. This is because plaster absorbs water from the air or floor, causing it to set in the bag (air set). Plaster should be stored:

* in the unopened bag it came in

* in an enclosed, well-ventilated room or building

* off the floor, for example on a wooden pallet

* clear of the walls, which may be damp

* in stacks of no more than five

* covered with a sheet or tarpaulin.

Plaster, along with any other bagged materials such as cement, sand, lime, aggregates and pre-mixed renders, should be rotated when new stock comes in. This means putting the newest bags (those with the use-by date that is longest away) on the bottom of the stack or the back of the store. This should mean that old bags are used before new bags, so that none is wasted.

CASE STUDY

South Tyneside Homes

South Tyneside Council's Housing Company

It's worth becoming an apprentice

Josh Gray is an apprentice plasterer at South Tyneside Homes.

'I'm studying for a Level 1 in plastering. I go to college once a week and the rest of the time I work for South Tyneside Homes. I tried out bricklaying and plumbing before but didn't enjoy it so decided to give plastering a go, as it was hands-on and sounded interesting. I looked online for apprenticeships and saw this one. I had to go in for a meeting and do a test and they took me on.

At college, you're improving the skills you learn in the workplace – you practise things like skimming and patching and putting on beads over and over, so when you go back to the workplace you're better at it. You also learn some of the theory, like the best plaster to put on different backgrounds.

I'm thoroughly enjoying it – you get paid well and you're learning new skills. I'd tell anyone thinking of doing an apprenticeship to go for it – it's one of the best things you can do.'

The limitations of materials relating to plasterboard fixing and plastering

We have seen in this chapter that the right materials need to be used in order to get the best result, and we have covered some of the issues you are likely to encounter on site. You need to take into account things like the finish required, the type of background material, the age of the building, dampness and humidity, time available and materials available.

For example, you may find that you are required to fix plasterboard to unsuitable, loose, painted or damp backgrounds. In this situation, you should not go ahead with the job; instead you should talk to your client or supervisor immediately, explaining the problem and thinking about how you can solve it. It's better for work to be delayed or new materials ordered than for the job to be completed knowing it's wrong.

Every job is different and, with experience, you will find ways of solving any problems you come across.

PRACTICAL TASK

1. PLUMB, DOT AND SCREED A SOLID BRICK WALL

OBJECTIVE

To use the plumb, dot and screed method to produce a floating coat that is level, flat and accurate to within 3 mm in a 1.8 m straight edge.

INTRODUCTION

On site, the use of dots and screeds is becoming less popular in favour of the dot and dab method. But this process is an important part of learning how to float and set a wall to the required national standards.

TOOLS AND EQUIPMENT

Buckets	Hop-up
Devil float	Plastering trowel
Dots (timber laths)	Spirit level
Flat brush	Spot board
Gauge	Straight edge
Handboard/hawk	

PPE

Ensure you select PPE appropriate to the job and site where you are working. Refer to the PPE section of Chapter 1.

STEP 1 Soak the timber laths in water and mix the mortar.

STEP 2 Place a small amount of mortar at the top right-hand side of the wall about 150 mm from the ceiling – enough to hold your dot in place. Place a pre-soaked timber lath on the mortar, allowing the lath to stick out about 12 mm from the wall. This is your first dot. Place another lath in a similar position on the left-hand side of the wall.

Figure 5.65 Placing the first dot

STEP 3 Directly underneath each dot, place another dot approximately 150 mm from the floor.

STEP 4 Using a straight edge and spirit level, adjust the dots and ensure they are plumb and in line.

Figure 5.66 The first four dots in place

STEP 5 Use a gauge to make sure the dots are set into the plaster at the required thickness and clean off any excess material

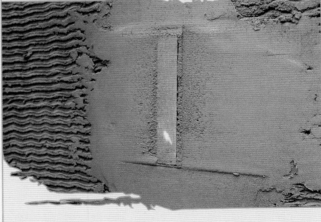

Figure 5.67 A dot flush to the wall

STEP 6 Check the dots are sufficiently firm to receive pressure from a trowel. This will take about 20 minutes.

STEP 7 Apply mortar horizontally between the dots, using the dots as a guide. Fill in any hollows to leave two bands of screed ruled out flat and level with each other.

Figure 5.68 Filling in between the dots

PRACTICAL TIP

To quicken the process and for training purposes you can add some casting plaster to the material when placing the dots. This will ensure you can rule off without damaging the dots.

STEP 8 Now rub up the screeds using a devil float, filling in any hollows or misses.

Figure 5.69 The screeds in place

STEP 9 Starting from the right-hand side of the wall (or the left if you are left-handed), and using the screeds as a guide, fill in with plaster. Rule off the excess material and check the wall for straight and plumb.

Figure 5.70 Filling in with plaster

Figure 5.71 Ruling off the screed

STEP 10 When the wall has set, use a devil float to key all areas of the surface to leave a suitable background to receive the finishing plaster.

Figure 5.72 Using a devil float

Figure 5.75 The finished wall

PRACTICAL TIP

Remove the dots and fill in any hollows before applying a setting coat to the wall.

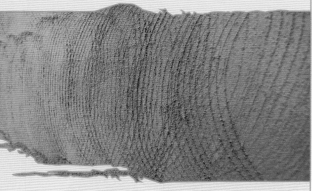

Figure 5.73 Close up of keyed surface

Figure 5.74 The difference in the textures

PRACTICAL TASK

2. APPLY A SETTING COAT TO A FLOATED WALL

OBJECTIVE

To complete a wall with finish plaster applied to the correct thickness and trowelled smooth.

INTRODUCTION

This task introduces you to the methods of applying setting coat plasters to a previously floated wall, using various types of finishing plasters.

Figure 5.76 Types of finishing plaster

TOOLS AND EQUIPMENT

Buckets

Flat brush

Handboard/hawk

Hand or mechanical whisk

Hop-up

Plastering trowel

Spot board

PPE

Ensure you select PPE appropriate to the job and site conditions where you are working. Refer to the PPE section of Chapter 1.

STEP 1 Check that the floated wall is flat and plumb and has sufficient key.

PRACTICAL TIP

Sometimes you will find hollows on the floated wall that will require filling before you apply the finishing coat.

STEP 2 Half fill a clean bucket with clean cold water and add the finishing powder slowly.

STEP 3 Stir thoroughly using a hand or mechanical whisk and ensure all the powder has been mixed in, with no lumps.

Figure 5.77 Mixing the plaster

STEP 4 Before you start plastering the wall, clean all the equipment by scrubbing it using a flat brush in a bucket of clean water.

STEP 5 Test the consistency of the mix and place enough on the spot board to work with without having to keep refilling it. Transfer the plaster from the spot board to the handboard/hawker when you need it.

STEP 6 Begin applying the plaster from the top left-hand side of your wall and work from left to right. Reverse this if you are left-handed.

Figure 5.78 Applying the plaster

STEP 7 Apply an even first coat to the entire wall, approximately 2 mm–3 mm thick.

Figure 5.79 The first coat

STEP 8 Allow time for the plaster to dry before applying the laying down coat (second coat of finishing plaster) to an even thickness. Again start at the top left-hand side of the wall.

PRACTICAL TIP

Left-handed plasterers start from the top right-hand side of the wall.

Figure 5.80 Applying the laying down coat

STEP 9 Start the first trowel of the wall (some plasterers call this the flatten trowel). Apply a little water at this stage.

Figure 5.81 The first trowel of the wall

Figure 5.82 Flicking water onto the wall

STEP 10 Start the second trowel of the wall. You should have no hollows or trowel marks at this stage.

Figure 5.83 Smoothing off

STEP 11 Now apply the final trowel of the wall. You can add some water with a flat brush to help you smooth the wall so that trowel lines won't show as the plaster starts to set.

Figure 5.84 The final trowel of the wall

STEP 12 Clean all the tools and equipment.

PRACTICAL TIP

Often colleges receive newly developed tools to test out. Take the opportunity to use them, and feed back your opinions to the manufacturer. The photo shows a Speedskim, a flexible rule that enables you to cover a large area of wall at once.

Figure 5.85 Ruling off with a Speedskim

PRACTICAL TASK

3. APPLY A SETTING COAT OF PLASTER USING PVA

OBJECTIVE

To complete a wall by applying finish plaster to a wall coated in PVA, ensuring it is the correct thickness and trowelled smooth.

INTRODUCTION

This task introduces you to the methods of applying setting coat plasters to a wall coated in PVA, using various types of finishing plasters.

TOOLS AND EQUIPMENT

Buckets	Mixing drill
Finishing plaster	Plastering trowels
Flat brush	PVA
Handboard/hawk	Spot board
Hop-up	

PPE

Ensure you select PPE appropriate to the job and site conditions where you are working. Refer to the PPE section of Chapter 1.

STEP 1 First ensure that the background is free from dust, is clean and sound, and that none of the existing plaster is loose.

STEP 2 Mix the PVA, diluting it to the manufacturer's recommendations in a clean bucket.

STEP 3 Brush on the PVA using a flat brush or roller. Start at the top of the wall and make sure you cover the entire area.

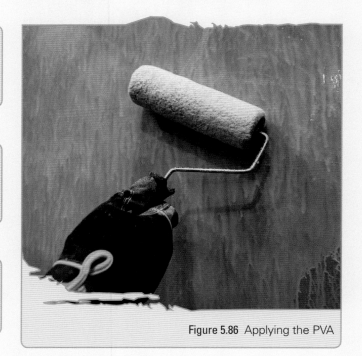

Figure 5.86 Applying the PVA

STEP 4 While the PVA is still tacky, apply the plaster by following steps 2–10 in Practical Task 2 above.

PRACTICAL TIP

Some painted surfaces can be skimmed after applying PVA, but you must always check the soundness of the background first.

PRACTICAL TIP

Using a textured and coloured bonding agent makes it easier to see if you have missed anything and gives an extra key before you apply the plaster.

STEP 5 Apply the final trowel to the plastered surface.

Figure 5.87 Applying the final trowel

PRACTICAL TASK

4. APPLY TWO COAT WORK USING LIGHTWEIGHT PLASTER

OBJECTIVE

To apply an undercoat of lightweight plaster to a solid brick wall followed by a finishing coat.

INTRODUCTION

The majority of two coat plastering involves using a type of lightweight undercoat plaster and covers a large range of backgrounds. Working to wet screeds is a skill all good plasterers need to learn.

TOOLS AND EQUIPMENT

Buckets	Hop-up
Darby	Mixing drill
Devil float	Plastering trowel
Flat brush	Spot board
Handboard/hawk	Straight edge

PPE

Ensure you select PPE appropriate to the job and site conditions where you are working. Refer to the PPE section of Chapter 1.

Figure 5.88 Types of undercoat plasters

STEP 1 Mix the undercoat by adding the powder to half a bucket of water. Ensure the mix is the right consistency before you use it.

STEP 2 Working from the right-hand side of the wall, apply a strip of plaster from wall to ceiling, approximately 11 mm thick. This is your wet screed.

Figure 5.90 Ruling off the undercoat

Figure 5.89 Applying the first wet screed to the wall

PRACTICAL TIP

Some plasterers tighten in (smooth) the floating coat, filling in the hollows with a darby and smooth off the plaster ready for the devil float.

STEP 3 Rule in the material with a straight edge and fill in any hollows.

STEP 4 Carry out the same process on the left-hand side of the wall and then horizontally at the top and bottom of the wall.

STEP 5 Now carefully fill in the middle of the wall with the undercoat plaster and rule in from all the screeds.

STEP 6 Now rub up with a devil float and fill in any hollows. Clean all angles with a trowel.

Figure 5.91 Wall keyed with a devil float

STEP 7 Carry out the skimming process by following steps 2–10 in Practical Task 2 above.

PRACTICAL TASK

5. FIX PLASTERBOARD TO A SMALL STUDDED WALL OR CEILING

OBJECTIVE

To practise your skills in measuring, cutting and fixing plasterboard, leaving a background surface that is ready to receive an application of finishing plaster.

INTRODUCTION

Plasterboard is produced in a many sizes and types and can be fixed to a variety of backgrounds using nails, screws or drywall adhesive. For this task you will use basic 1,200 mm × 900 mm plasterboard and fix it with drywall screws.

TOOLS AND EQUIPMENT

Drill	Saw
Foot lifter	Spirit level
Hammer	Straight edge
Hop-up	Tape measure
Pad saw	Trimming knife

PPE

Ensure you select PPE appropriate to the job and site conditions where you are working. Refer to the PPE section of Chapter 1.

STEP 1 Firstly check the size and thickness of the plasterboard specified for the job.

STEP 2 Check the timber studding joists are in line, level and at the correct centres.

STEP 3 Plan the layout of the plasterboards to ensure the joints are staggered across the wall.

STEP 4 Measure for the first piece of plasterboard, working from the bottom right of the wall in order to fix the board horizontally across the joists.

PRACTICAL TIP

Always measure twice and cut once to avoid mistakes, which can be costly.

STEP 5 Lay the board flat and mark off the measurements required. Place the straight edge up to these marks and, using the trimming knife, score the board two or three times along the straight edge. Then turn the board over and tap along the back of the cut until it breaks. Once the board is broken, cut the paper lining neatly with the trimming knife.

Figure 5.92 Cutting the plasterboard

STEP 6 Position the plasterboard, using a foot lifter or the help of a colleague to place it without straining your back and arms. Starting from the edge, fix drywall screws at 300 mm centres.

Figure 5.93 Snapping the plasterboard

Figure 5.94 Safely lifting the plasterboard sideways

Figure 5.95 Safely lifting plasterboard above head height

Figure 5.96 Fixing the plasterboard

STEP 7 Make sure that the head of the screw just grips the paper of the plasterboard, without puncturing it.

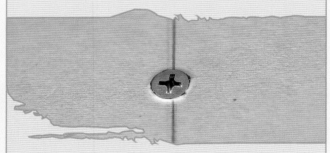

Figure 5.97 The screw head positioned correctly

STEP 8 Measure, cut and fix the next piece of plasterboard following the same procedure. Make sure the end joints are staggered to prevent continuous cracking and a 2 mm–4 mm gap is left between each board.

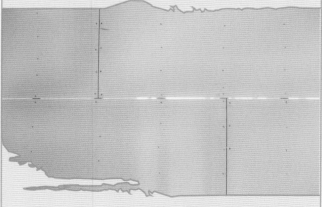

Figure 5.98 Staggering the plasterboard

STEP 9 Continue until you have covered the area. When fixing the plasterboards at height, ensure you use the correct scaffold or hop-up.

STEP 10 Fix the fibre tape to each of the plasterboard joints. If you are working to brickwork or other plasterboard, apply it around all the edges, keeping the tape flat and even across all the joints.

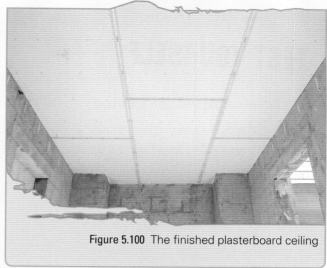

Figure 5.100 The finished plasterboard ceiling

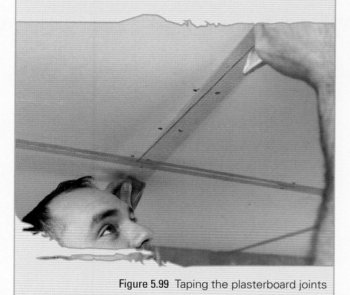

Figure 5.99 Taping the plasterboard joints

TEST YOURSELF

1. When would you use a darby?

   a. To provide a key in backgrounds

   b. To straighten undercoat on walls and ceilings

   c. To measure the distance from wall to ceiling

   d. To get a sharp right-angle

2. What would you use to get a sharp edge or corner?

   a. EML

   b. Scrim tape

   c. Beading

   d. A gauging trowel

3. What is a floating coat?

   a. An undercoat applied before the setting coat

   b. The coat applied on top of the setting coat

   c. A coat applied using all-in-one plaster

   d. A thin finish or setting coat

4. Why do you put new bags of plaster behind or underneath older bags?

   a. To stop them getting damp

   b. So that they absorb any water in the storage area

   c. So that the older bags are used first

   d. Because they are worth more so need to be kept secure

5. What is an expansion joint?

   a. Cracking in the plaster as it dries

   b. A tool to create clear divisions between types of materials in a structure or wall

   c. A break between sections of a structure or wall to allow for the construction materials to move

   d. Pipework that creates a hazard when plastering

6. What is EML?

   a. An all-purpose plaster

   b. A break between two different types of material

   c. A type of risk assessment

   d. A metal mesh that provides a key

7. In lightweight plasters, aggregates are used instead of which material?

   a. Sand

   b. Cement

   c. Lime

   d. Gypsum

8. What is the main advantage of using the broad screed method instead of the plumb, dot and screed method?

   a. It produces a better finish

   b. It is quicker

   c. It is more accurate

   d. It uses modern techniques

9. Where are you most likely to use lime plaster?

   a. In new-build houses

   b. In damp environments

   c. In refurbishment projects

   d. In long stretches of corridor

10. What is another name for two coat work?

    a. Broad screeding

    b. Float and set

    c. Render, float and set

    d. Skimming

Chapter 6

Unit CSA–L10cc18
PRODUCE COMPONENTS FROM MOULDS

LEARNING OUTCOMES

LO1/2: Know how to and be able to prepare for producing components from moulds

LO3/4: Know how to and be able to produce plasterwork components from moulds

INTRODUCTION

The aims of this chapter are to:

* help you to understand the purpose of producing components from moulds

* show you how to create a running mould and a flood mould

* explain the effects of different moulding materials and additives

* show you how to create components from different types of mould

* explain how to protect components.

Fibrous plastering is a specialist area of the plastering trade and involves producing ornamental components from **moulds**. It also requires different tools than those you would use when working with solid plastering, and is a very different skill to master. Plasterers who specialise in producing moulded components are based in workshops and rarely, if ever, carry out other aspects of plastering, such as rendering or dry lining. However, it's worthwhile learning the basic skills, even if you don't go on to become a fibrous plasterer.

Components are **cast** from moulds off site, or at least away from where they are going to be used, before being fixed into position.

PREPARING TO PRODUCE COMPONENTS FROM MOULDS

Controlling hazards when producing components from moulds

Casting plaster (plaster of Paris), as with standard plaster, is not dangerous if worked responsibly. However, plaster of Paris is classified by the Health and Safety Executive (HSE) as a hazardous substance so a risk assessment is required by law before it is handled.

Normally you will be casting using fairly small quantities of plaster. However, you still need to take care, because casting plasters give off a fine dust and can also irritate skin. Plaster of Paris is often used to take casts from body parts like arms or heads. While it's unlikely that you will be asked to do this, don't assume that it's safe to use on bare skin. Plaster of Paris can heat up to 60°C or more, and temperatures of just 45°C can burn. It's worth taking the following precautions:

* Have a bucket of cold, clean water, a sponge and a towel available in case you need to rinse plaster splashes off your skin.

* When you have finished casting, clear up thoroughly and make sure no plaster dust remains that could cause breathing difficulties.

* Never pour wet plaster down the sink, even if it's only a small quantity. Ask your employer, client or site manager how to dispose of plaster materials.

* Place all plaster fragments in a rubbish bag and seek advice about how to dispose of it.

You may also use additives and chemicals in the moulding and casting process. Many of these are toxic so, as with plaster, you will need to control their use, for example by using only the quantities you need.

PPE when producing components from moulds

The risk assessment should outline actions to reduce the chance of injury. However, you should still wear appropriate PPE, such as overalls, goggles, gloves and a dust mask (if mixing dry powder indoors or using chemicals). Ensure these are clean before use and are cleaned after use.

The purpose of plaster moulds

A mould is like a template – it forms the basis for the decorative plasterwork you will be producing. Moulds are used to produce cornices, dado rails, corbels, columns, arches and coving, as well as more complex features like ceiling roses. Although you can buy these sorts of casts off-the-shelf at builders' merchants and DIY stores, individual items created from moulds can not only look more attractive but also enable the production of unique designs to fit the area where the decorative plasterwork is going to be fixed. It is much easier to create these items off-site than to try to make them where they are to be located, which may be a high ceiling or wall.

Figure 6.1 Detail of the ceiling decorations at Blickling Hall

DID YOU KNOW?

Plaster mouldings are commonly seen in historic buildings. A particularly intricate example is in the Long Gallery at Blickling Hall, a National Trust house in Norfolk. At nearly 40 m long, it includes intricate geometric patterns and images of people.

Tools used for producing positive moulds

As well as your normal tools, like a trimming knife and measuring equipment, particular tools are needed to produce moulds. Table 6.1 describes the main ones.

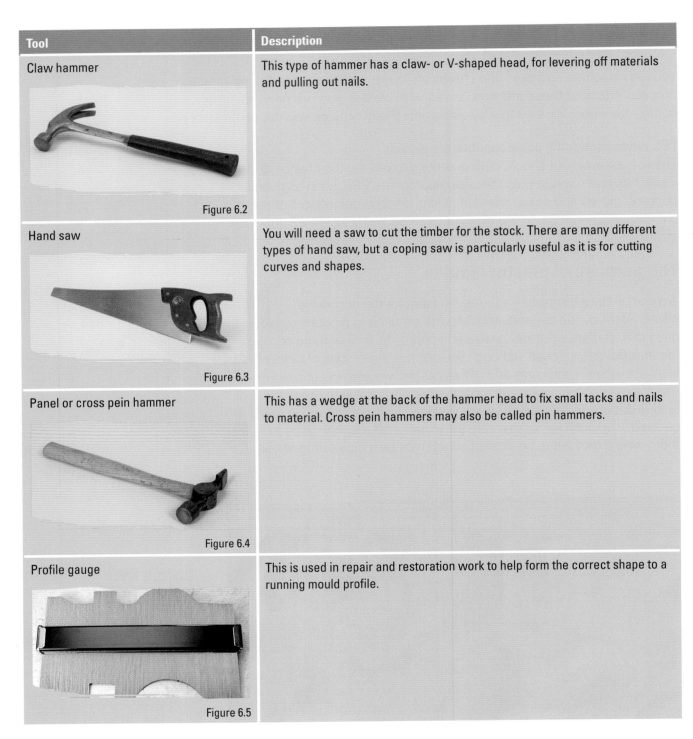

Tool	Description
Claw hammer Figure 6.2	This type of hammer has a claw- or V-shaped head, for levering off materials and pulling out nails.
Hand saw Figure 6.3	You will need a saw to cut the timber for the stock. There are many different types of hand saw, but a coping saw is particularly useful as it is for cutting curves and shapes.
Panel or cross pein hammer Figure 6.4	This has a wedge at the back of the hammer head to fix small tacks and nails to material. Cross pein hammers may also be called pin hammers.
Profile gauge Figure 6.5	This is used in repair and restoration work to help form the correct shape to a running mould profile.

Tool	Description
Rasp  Figure 6.6	Also known as a planer file, this is a long steel stick with a handle. Its rough, spiky surface shaves off irregularities and smooths down the plaster. It is used to form profiles or templates, and to produce running moulds.
Small tool Figure 6.7	Despite its generic name, this is a tool in its own right. It is a flexible steel hand tool most commonly available in two designs: the leaf and square, and the trowel and square, and each design comes in three sizes of between 11 mm and 25 mm. Small tools are used when small mixes are required and where there is small detailed work is to be completed. Trowel and squares are also excellent for stopping in (caulking), and leaf and squares are used for measuring out semi-viscous fluids like silicones and finishing off cast items.
Tin snips Figure 6.8	This has long handles and short blades with extra-wide jaws, and is used for cutting soft metal, such as beading or EML (see Chapter 5, page 156). Straight pattern tin snips cut straight lines or gentle curves, while duckbill or Trojan-pattern snips cut larger curves and circular shapes.
Square Figure 6.9	This is a metal or plastic tool that provides a 90° angle to help you check for square. It may also have measurements on it like a ruler. Other drawing tools, such as pencils and a compass, might also come in handy when constructing moulds.
Surform® Figure 6.10	This is used to take high points off the timber slipper before the running mould is constructed. It is a flat tool with a base like a cheese grater, with open teeth that allow shavings to pass through.

Table 6.1 Tools for making moulds

Constructing a running mould

A running mould is a wooden and metal structure used to create a straight or circular panel or dado mouldings. The mould is the template and the plaster is poured into it. Once the plaster has set, it is fixed in place.

To create a panel moulding, you must first create the running mould and then use the mould to shape the plaster. Running moulds are made up of two parts:

1. a template of the shaping being formed

2. a wooden frame (stock) for the template.

Making a running mould template

To get an accurate tracing of the moulding you want to copy, you either have to remove a section and trace round it, or make a saw cut across the profile, insert stiff paper or card into the cut, and draw around it.

Draw the profile you want to use on a piece of paper and glue it onto a plate of zinc. Using snips, carefully cut the zinc around the profile line and file it to make a clean edge. Check it against the drawing.

Lay the profile on the stock and draw around it with a pencil, then draw it again about 5 mm further out. Saw the wood to this second line to allow for swelling.

DID YOU KNOW?

For some very large projects, steel is used instead of zinc. As it's much stronger than zinc, it is more sturdy, will last longer and will not bend. However, it's also much harder to cut. Steel running moulds would normally be cast in situ on a wall or ceiling and the resulting cast may need coring out with heavy materials such as sand or cement.

The running mould frame

The frame is constructed from plywood or timber with two straight edges. It is made of three parts:

* A stock – a support for the template, to prevent it from bending during the running.

* A horse or slipper – a support for the stock, which is fixed to it at a right angle (90°). The horse runs along the bench's running rule, and is 1.5 times as long as the stock.

* A brace – a support for the stock and horse to keep the frame from slipping. The plaster swells as it sets, so may try to push the frame out of alignment. Larger moulds may need two braces to minimise movement.

If you are producing a circular moulding, you also need a piece of wood known as a gig stick (see Fig 6.12). There are two ways of using this:

1. The gig stick is slightly longer than the radius of the circle you need, and you nail it flat on top of the stock and hand brace. Then the stick is nailed to the bench table at the other end so that it can go round in a circle.

2. The gig stick forms an extended stock, so needs to be long enough to extend from the slipper to beyond the turning point. Then, to allow the running mould to rotate, you need to fix a turning eye at the end of the gig stick. This is a piece of metal with a 50 mm hole in it. Then a turning block is nailed below the turning eye to attach the mould to the bench. It can be further held in place with a piece of canvas soaked in plaster (a wad).

You are now ready to run out the mould.

PRACTICAL TIP

Splaying the outline's edge away from the template stops the plaster from clogging it up.

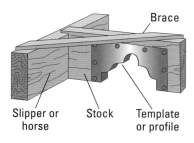

Figure 6.11 Example of a running mould of a simple panel moulding

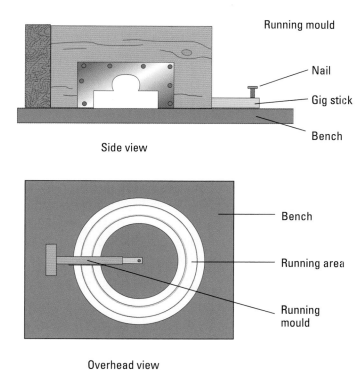

Figure 6.12 A circular running mould using a gig stick

Materials used in producing components from moulds

Casting plaster

As ordinary plaster is not fine enough to produce sufficient detail, you will also need casting plaster. This is also known as plaster of Paris or Class A hemi-hydrate plaster. This has no retarder added and tends to set very quickly. It is normally available in bags of 1 kg to 25 kg. Table 6.2 describes some different types.

Type of casting plaster	Description
Super fine	As its name suggests, this is extremely fine, so that you can produce very detailed patterns. It can also be used in combination with fine casting plaster.
Fine	This is the standard casting plaster, which is soft enough to be carved, sanded and shaped. British Gypsum fine casting plaster takes 18–22 minutes to set.
Coarse	This is usually used as a cheap way of coring out moulds (producing the inside part) but is not usually used where it can be seen because its finish is not as good as for fine or super fine plaster.

Table 6.2 Types of casting plaster

Brands to look out for are Teknicast, Helix, Herculite, Crystacal R and Crystacast. The two last types are extremely hard and dense. Each type of plaster has its own characteristics and, with experience, you can achieve excellent results.

Additives

Numerous additives can be added to the basic plaster mix to change its properties, or used separately to make the moulding process easier.

Reinforcement

Moulds made from plaster of Paris are fragile and so may need to be reinforced to ensure they last as long as they are intended to. If the mould is made off site, it needs to reach its destination in one piece. Other than using cement-based plasters (as discussed above), plaster can be strengthened in a number of ways:

* By incorporating a scrim of hessian or canvas. This is most appropriate for large mouldings.

Figure 6.13 Using hessian scrim

- By adding chopped glass fibre strands to the mix or, for larger or less detailed moulds, incorporating fibreglass mats into the cast.

- By adding PVA adhesive to the mix.

Release agents

Release agents stop the cast from sticking to the mould, and they also act as a barrier between the casting media and the mould, to help prolong the life of the mould.

- Plasterers commonly keep a stock of grease to use as a release agent. Most casting shops mix it up themselves using tallow and paraffin or engine oil.

- A more expensive but less labour-intensive alternative is ready-made petroleum jelly (such as Vaseline) but this may inhibit the setting of glass fibre resin.

Figure 6.14 Fibreglass reinforcement

Retarders

Casting plasters tend to set in 15 to 20 minutes, which is often too quick to complete the cast. Retarders are used to lengthen the setting time. If you are making cornices, it means that you could prepare two plaster mixes at once and complete the job in one go.

- Traditionally, plasterers use glue size, which is a jelly-like material. The crystals are stirred bit by bit into very hot water until they have dissolved. Then hydrated lime is added to prevent the mixture from becoming solid.

- A less smelly alternative is trisodium citrate. Again, you can add a pinch of the crystals to water, or directly into the plaster mix if you are confident about how much you need. You don't need to add lime.

PRACTICAL TIP

Neither option has a recognised mixing ratio, as the level of retardation depends on the individual job. However, it's worth mixing up a test pot, as using too much glue size could considerably lengthen the setting time, while too much trisodium citrate will prevent the plaster from setting at all.

Sealing compound

You must wait until the plaster is completely dry before sealing it. Shellac is most commonly used for this: it is a painter's polish or varnish that seals the plaster surface and makes it waterproof. You dilute with methylated spirit at a ratio of 1:3 or 1:4 and then apply it in three or four coats.

PRACTICAL TIP

As well as adding to the overall cost of the plasterwork, it is worth remembering that additives may affect the plaster in unexpected ways – changing its chemical make-up may prevent it setting or stain the plaster if not used properly. As always, check the manufacturer's data sheet before using them.

PRACTICAL TIP

Always use appropriate PPE, such as gloves, goggles and dust masks when using casting plaster and its additives. The chemical fumes in particular can be toxic, so it may be necessary to use a respirator or other barrier against breathing-in the fumes.

Protecting the work and the work area from damage

If you are on a large site with other trades, it's best to carry out moulding work in a separate workshop. If this isn't possible, you will need to agree an appropriate space with the site supervisor where people are not going to interrupt or accidentally damage the mouldings as they pass.

Put away tools and moulds when you leave site, to prevent accidents, theft and damage. The store should be secure and watertight.

Forming flood moulds and casts

KEY TERMS

Model

– the original item that is being reproduced.

In addition to using a panel running mould, you can use different methods to form moulds from a **model**. For example:

* a flood mould (also known as solid or open mould): made by pouring moulding material over an open mould

* a case mould (also known as clay case mould or closed mould): made by casting the mould from a model

* a skin mould: made by building up a layers of compound over the model until they form a thin skin.

How you carry out the method you choose depends on:

* how much time you have

* how much material you have

* how much money is available

* the equipment that is available.

This section explains how to form a flood mould.

When forming a flood mould, you can use either cold pour moulding compound or hot melt compound (PVC).

Cold pour moulding compound

This is a flexible silicone rubber that, as its name suggests, does not need to be heated before use. It must be stored in a sealed container (a blast storage bin). This is because it consists of a chemical and a catalyst that you must mix together to begin the setting process.

Figure 6.15 Cold pour moulding compound

You can buy different grades of cold pour compound to suit the work you are doing. Table 6.3 shows the pros and cons of using this method.

Advantages	Disadvantages
The mould will be very strong and flexible	It cannot be reused so is expensive
It will provide accurate and detailed reproduction of the original model	It often takes experience to accurately measure quantities for the right mix
It is easy to prepare the cold pour on site	The chemicals used are toxic and flammable (may catch fire)
	Some cold pours may not set properly if they are contaminated with other things
	Both parts must be thoroughly mixed to obtain a uniform cure/set

Table 6.3 Advantages and disadvantages of cold pour moulding compound

Hot melt compound (HMC)

Hot melt compound is a rubbery material known as polyvinyl chloride (PVC). It is a thermoplastic vinyl resin that comes in a variety of colour-coded grades depending on the flexibility you require.

To melt the compound, you need to use an electrical thermostatically controlled heating machine (see Fig 6.61). PVC melts at 120–170°C, depending upon the grade chosen.

Table 6.4 shows the pros and cons of using this method.

Advantages	Disadvantages
Reproductions are of a reasonable quality	If the PVC is too hot, the model may release air bubbles that will affect the finish of the final cast
The mould remains strong and flexible unless it is melted down too often	Hot melt machines are expensive, so the initial outlay may be more than small businesses or individuals can afford
You don't need to mix up the materials and measure quantities so less is wasted	Working with hot materials that produce fumes presents safety hazards
The mould can be cut up and reused when you have completed the job it was made for	

Table 6.4 Advantages and disadvantages of hot melt compound

Figure 6.16 Hot melt compound

PRODUCING PLASTERWORK COMPONENTS FROM MOULDS

Table 6.5 shows the equipment you will need.

Equipment	Description
Bench  Figure 6.17	The bench needs to be big enough to run the mould you are making, ideally about 3 m × 1 m or a metre square. It should be made of any strong timber and have a plaster or laminate (kitchen) top with a wooden or metal rule on each side. The top needs to be greased to stop the plaster sticking and to help the mould break free of the wood. You won't need your own bench when you're starting out, but you should check that the right type of bench is available on site. It is possible to make your own from plywood.
Bowls  Figure 6.18	Small bowls are useful for mixing up small batches of plaster. These may be made from flexible, heavy duty rubber or polythene, so that you can easily press out dried plaster. Typically they have a diameter of 250 mm.
Buckets  Figure 6.19	Heavy-duty buckets are necessary not only to mix plaster but also to carry materials, water and waste. Useful sizes are between 25 and 65 litres.
Canvas or hessian  Figure 6.20	This is used as a standard reinforcement for plaster casts. It usually comes on a roll of several metres of material, which can vary in width from 75 mm up to 900 mm. It has a 3 mm or 6 mm square mesh jute weave which can be cut using scissors to any length required.

Equipment	Description
French chalk 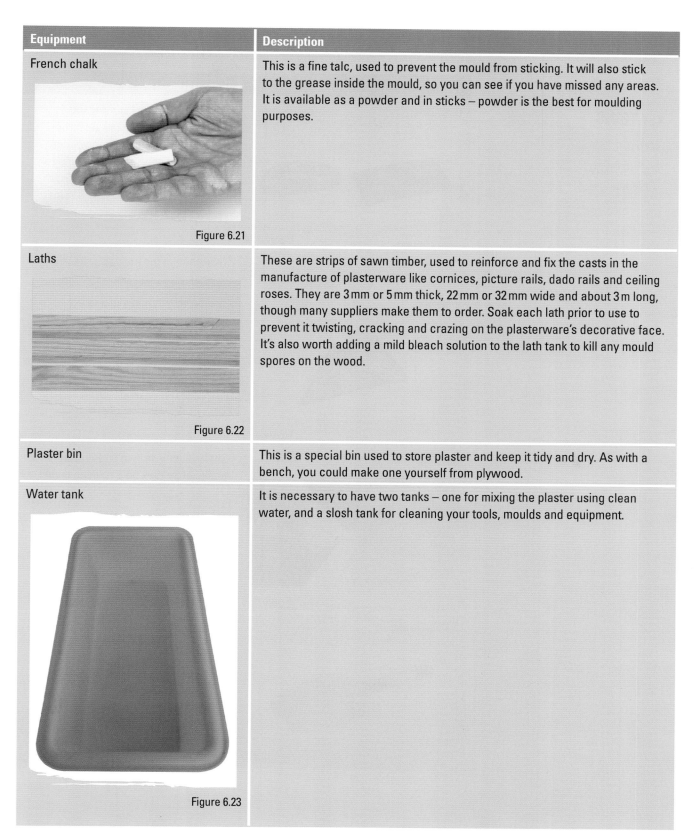 Figure 6.21	This is a fine talc, used to prevent the mould from sticking. It will also stick to the grease inside the mould, so you can see if you have missed any areas. It is available as a powder and in sticks – powder is the best for moulding purposes.
Laths Figure 6.22	These are strips of sawn timber, used to reinforce and fix the casts in the manufacture of plasterware like cornices, picture rails, dado rails and ceiling roses. They are 3 mm or 5 mm thick, 22 mm or 32 mm wide and about 3 m long, though many suppliers make them to order. Soak each lath prior to use to prevent it twisting, cracking and crazing on the plasterware's decorative face. It's also worth adding a mild bleach solution to the lath tank to kill any mould spores on the wood.
Plaster bin	This is a special bin used to store plaster and keep it tidy and dry. As with a bench, you could make one yourself from plywood.
Water tank Figure 6.23	It is necessary to have two tanks – one for mixing the plaster using clean water, and a slosh tank for cleaning your tools, moulds and equipment.

Table 6.5 Equipment for casting plasterwork

As well as your standard tools (such as hammers, saws and small tools), you will need some specialist tools to help you make your moulding.

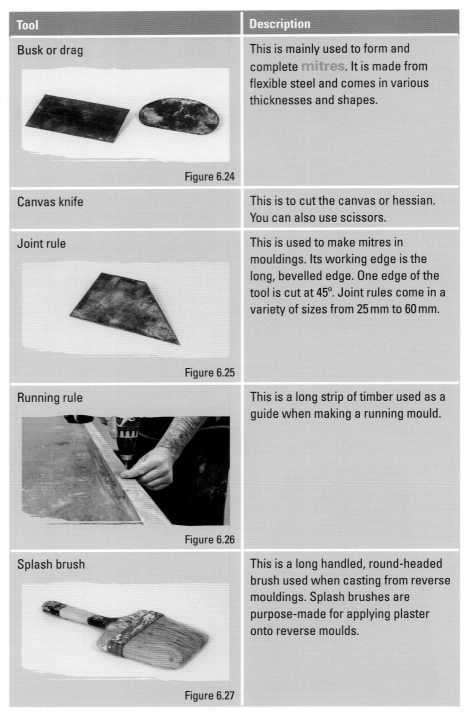

Tool	Description
Busk or drag Figure 6.24	This is mainly used to form and complete mitres. It is made from flexible steel and comes in various thicknesses and shapes.
Canvas knife	This is to cut the canvas or hessian. You can also use scissors.
Joint rule Figure 6.25	This is used to make mitres in mouldings. Its working edge is the long, bevelled edge. One edge of the tool is cut at 45°. Joint rules come in a variety of sizes from 25 mm to 60 mm.
Running rule Figure 6.26	This is a long strip of timber used as a guide when making a running mould.
Splash brush Figure 6.27	This is a long handled, round-headed brush used when casting from reverse mouldings. Splash brushes are purpose-made for applying plaster onto reverse moulds.

Table 6.6 Tools used for casting

CASE STUDY

Look after your tools

Glen Campbell, Property Services Team Leader at South Tyneside Homes says:

'If you don't look after your gear, it won't look after you. If you don't keep it sharp, you're going to hurt yourself because you've got to apply more pressure to get the job done. This includes everything from sharpening the bits for your drills, making sure your power tools aren't clogging up or jamming and that they're oiled, and oiling your blades. Don't go putting your gear away in a dusty place, otherwise your motor will just get clogged up again.

Always use the right bit for the right tool. This is so important. The newer guys don't always realise… they'll say, 'Ooh, that's only two pounds for that bit, and I paid 15 before!' Well, there's a reason why you paid £15 for that part.

Looking after your gear also means maintaining it and making sure it's PAT tested. It all comes back to health and safety really. You absolutely have to use the right PPE otherwise you're likely to end up with something in your eye, or hurting your hands when you should have been wearing gloves.'

Methods of casting and running flood moulds

To form a flood mould, you can use casting plaster or a moulding compound, depending on what the mould will be used for. You are simply pouring the material over the model in an open mould, usually a wooden or clay frame, to reproduce it.

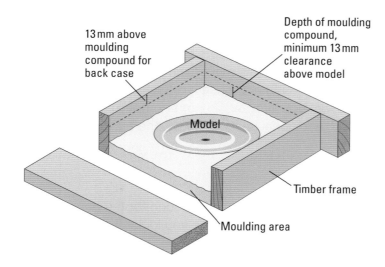

13 mm above moulding compound for back case

Depth of moulding compound, minimum 13 mm clearance above model

Model

Timber frame

Moulding area

Figure 6.28 A flood or open mould

To run a flood mould using plaster, you can use either the two-gauge or the one-gauge (sometimes called single-gauge) casting system.

Two-gauge casting system

This uses two plaster mixes, called firstings and seconds. As you might guess, the firstings are mixed and applied first, and the seconds are applied along with the canvas after the firstings have set. The seconds are mixed with size to slow the set. The idea is that a second layer gives a better finish because otherwise the canvas might show through on the face.

One-gauge casting system

This uses a single mix of plaster with size. This system avoids two mixes swelling at different rates but has the disadvantage of possibly pressing the canvas through to the face.

See the practical tasks on pages 191 to 201 for details of how to make a flood mould and carry out each casting system.

Drying and storing plaster casts

Drying plaster casts

Plaster of Paris takes 15 to 30 minutes to set, though this may vary according to the brand and any additives used, and it may not completely set for several hours. The time taken will also depend on factors such as:

* humidity (the amount of water in the air)

* room temperature (it will set more quickly in a warm environment)

* water additives or mineral content

* the amount of water used

* the method of mixing.

Plaster doesn't really dry – it sets by the powder chemically bonding with water in the mix. This makes the plaster hot during the setting process so you should not touch it or attempt to move it until it has set. For this reason, you also should not attempt to speed up the setting process with additional heat. If you do, the plaster may explode.

It is best to let it set naturally, in a warm, well-ventilated room.

Storing plaster casts

Although casts should be strong enough for the purpose they were made, you still need to make sure you look after them by handling and storing them properly.

* Wait until they have set before attempting to move them.

* Always handle casts on their edge.

REED TIP

If you're having trouble, always ask for help. It's better to ask and find out than to blunder on and make mistakes. By getting the help you need, you'll learn the skills properly, you'll avoid wasting time and money, and you'll have a better chance of succeeding in your future career.

* Lift them carefully, with more than one person if they are heavy.

* Store casts away from workshop activity – in a storeroom if possible, or at least in a quiet corner.

* Store them flat or, if they are long (such as a cornice), tie them vertically against a scaffold board leaning against the wall, face inwards.

* Cover casts with a dust sheet or light tarpaulin to prevent them from becoming dirty or damaged.

It is best to time the making of casts so that they don't need to be stored for too long before they are fixed.

PRACTICAL TASK

1. CONSTRUCT A RUNNING MOULD

OBJECTIVE

To draw a moulding section from information provided using basic geometry, and then to construct a running mould by transferring the moulding outlines onto metal profiles using woodworking skills.

INTRODUCTION

Running moulds are made from timber and metal. They are used to form shapes in plaster. These plaster mouldings are run in neat plaster of Paris for maximum strength.

You can run straight or curved mouldings using different techniques.

TOOLS AND EQUIPMENT

Coping saw	Set square
Files	Tape measure
Nibblers	Tin snips
Hammer	Trimming knife
Profile paper	Vice
Scissors	Zinc sheet

PPE

Ensure you select PPE appropriate to the job and site conditions where you are working. Refer to the PPE section of Chapter 1.

DID YOU KNOW?

Nibblers are similar to tin snips, but have more of a shearing action that helps to cut metal without distortion.

STEP 1 Draw a basic profile onto paper, making it no more than 100 mm long and 50 mm high.

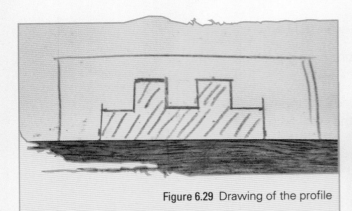

Figure 6.29 Drawing of the profile

STEP 2   Place the paper profile over a piece of zinc sheet and mark out the profile by drawing around it onto the zinc.

Figure 6.30 Marking out the profile

STEP 3   Use tin snips to cut out the basic shape and nibblers to cut out the profile on the zinc to approximately 2 mm outside the line.

Figure 6.31 Cutting the zinc with snips

STEP 4   Place the cut zinc in a vice and, using small square and round metal files, carefully file down to the line to form a clean cut zinc profile.

Figure 6.32 Filing the profile

PRACTICAL TIP

Using wire wool, clean off any edges on the zinc profile to leave a smooth, sharp, clean edge.

STEP 5   Saw the pieces of timber to length. For basic running moulds, the stock should be 100 mm high by 150 mm long. The horse/slipper should be one and half times the length of the stock (so it needs to be 225 mm if the stock is 150 mm long).

Figure 6.33 Cutting the timber for the stock

PRACTICAL TIP

Make sure you cut all timber square. Use sandpaper to give a smooth finish.

STEP 6 Now draw round the zinc profile onto the stock. Then draw a second line 5 mm away from the profile line.

STEP 7 Place the stock in the vice and use a coping saw to cut out the timber to the second line.

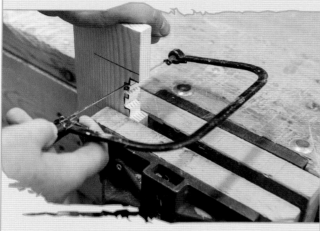

Figure 6.34 Cutting out the profile shape

STEP 8 Now fix the zinc profile into position onto the stock using panel pins.

Figure 6.35 Fixing the zinc profile to the stock

STEP 9 Measure to the centre of the horse/slipper lengthwise and mark it. Drill a hole through the point you have marked and screw the stock and slipper together.

Figure 6.36 Fixing the slipper to the stock

STEP 10 Cut a piece of timber to form a brace. This needs to be long enough to stretch from the far top edge of the stock to one end edge of the slipper. Make sure the timbers are square then fix the brace to the slipper and stock using screws or nails.

Figure 6.37 Fixing the brace to the stock

STEP 11 Sand down the running mould and ensure it is square and flat. It is now ready to run.

Figure 6.38 The finished running mould

PRACTICAL TASK

2. RUN A STRAIGHT SECTION OF MOULDING

OBJECTIVE

To mix the casting plaster and to produce a small mould such as a dado or panel mould, using the running mould you made in Practical Task 1.

INTRODUCTION

The running mould is run against a running rule to ensure that the moulding is run straight. You must make sure that the moulding is run on the bench and prevent any materials gathering underneath the running mould.

TOOLS AND EQUIPMENT

Casting plaster	Hammer
Drag or busks	Pin nails
Grease	Running rules
Joint rule	Small tools
Mixing bowls	Splash brush

PPE

Ensure you select PPE appropriate to the job and site conditions where you are working. Refer to the PPE section of Chapter 1.

STEP 1 Form a running rule by fixing a straight piece of timber about 1.2 m long to the bench using screws or nails.

Figure 6.39 Fixing the running rule

STEP 2 Mark out the length of moulding required on the running rule.

Figure 6.40 Marking out the length of moulding

STEP 3 Check the bench area is clean and free from hollows. If it is not flat, fill in the hollows with a little plaster and leave it to set before starting to run the mould.

STEP 4 To keep the mould in place, fix anchorage nails using the lengths marked on the running rule as a guide. Ensure that they are positioned in the thickest part of the moulding section but don't put one at either end, as they might snap the plaster as it swells. Check that the moulding does not catch on the nails before preparing the mix.

PRACTICAL TIP

To make it easy to release the plaster from the nail once you have completed the run, put a little pyramid of clay over the head of the nail.

Figure 6.43 Greasing the running mould

PRACTICAL TIP

It is especially important to grease the running mould around the zinc profile area, as this will help the running mould to run freely on the bench.

Figure 6.41 Nails being fixed and running rule in place

STEP 6 Mix a full bowl of plaster of Paris until it is the consistency of single cream. Use a small piece of lath or small tool to mix the plaster.

PRACTICAL TIP

The amount of plaster you need will depend on the size of the section of moulding. It takes some experience to get it right. If in doubt, mix a smaller amount but try to ensure that the bulk of the moulding is completed with the first mix.

STEP 5 Evenly grease the whole area where the plaster will be applied.

Figure 6.42 Greasing bench

Figure 6.44 Casting plaster mixed to the correct consistency

STEP 7 Pour the mix from the bowl onto the bench area. Place the running mould at the far end with the template facing along the length of the running area. With one hand on the hand brace, firmly pull the mould against the running rule as you walk backwards, keeping it flat on the bench. Catch the excess plaster in a bowl at the end of the bench so that you can use it again. As members (straight and curved shapes) build up, add more plaster.

Figure 6.45 Pouring the mix

Figure 6.46 Running the mould

STEP 8 If another mix is required, follow steps 6 and 7 using a smaller amount of plaster. If canvas is needed to reinforce the moulding, soak it in plaster before use. Make sure you wash the mould after each run and clean the edges of the mould with a joint rule. The mould is ready when it fits the profile of the running mould without any hollows.

Figure 6.47 Soaking the canvas in plaster

Figure 6.48 Using water to clean the edges of the mould

STEP 9 Once the moulding has fully set, cut it to length using a small-toothed saw and plenty of clean water to damp it down and reduce dust.

Figure 6.49 Cutting the moulding to length

STEP 10 Finally, lift the mould carefully from the bench using a joint rule and store it on a flat, dry surface, away from where it might be damaged. Ensure you remove the anchor nails from the bench.

Figure 6.50 The finished moulding

PRACTICAL TASK

3. PRODUCE A BASIC PLASTER CAST FROM A FLOOD MOULD

OBJECTIVE

To practise the skills necessary to produce a plain-faced slab in plaster, including preparing a bench surface and constructing and fixing the retaining walls (fence).

INTRODUCTION

This task will introduce you to basic casting methods and using reinforcement materials for the slab.

Make sure you are aware of the required finish for the face of the slab.

Mixing the materials in the correct manner is essential when casting.

TOOLS AND EQUIPMENT

Casting plaster	Pin nails
Drag or busks	Running rules
Grease	Shellac
Hessian	Small tools
Joint rule	Splash brush
Mixing bowls	Timber
Pin hammer	

PPE

Ensure you select PPE appropriate to the job and site conditions where you are working. Refer to the PPE section of Chapter 1.

STEP 1 Cut four pieces of timber running rule to a length of 350 mm each. Use screws to fix them together by the ends in a square to make a fence. Fix the fence to the bench.

PRACTICAL TIP

Soak the timbers overnight before use.

Figure 6.51 Running rules in place

STEP 2 Check that the surface of the bench is clean and fill in any hollows with plaster. Seal with shellac, and then apply the grease over the entire bench area.

STEP 3 Next, cut the hessian canvas into two squares that overlap the timber fence by 50 mm on all sides. These will be the facing and backing canvas. When they are the correct size, put them aside.

Figure 6.52 The canvas in position over the fence

STEP 4 You are now ready to prepare the firstings and seconds. Remember the seconds always have retarder added and so should be mixed first, so mix the seconds in a bowl, to the consistency of double cream and leave to soak (put it aside) to be used later.

STEP 5 Now mix the firstings. These should also be the consistency of double cream.

Figure 6.53 Mixing the firstings and seconds

STEP 6 Pour a small amount of the firstings and brush it over the whole bench area.

Figure 6.54 Brushing over the firstings

STEP 7 Now, holding the bristles of your splash brush, splash a layer of plaster up to 3 mm thick all over the mould.

Figure 6.55 Applying plaster to the mould

STEP 8 Position the facing canvas over the layer of plaster and, using your splash brush, soak this canvas with a layer of seconds.

Figure 6.56 Soaking the canvas in the seconds

STEP 9 Now position the backing canvas over the layer of seconds and cover it with another layer of plaster. Make sure you turn back the canvases so they are not proud of (overhanging) the strike-off points.

Figure 6.57 Turning back the canvas

STEP 10 When the cast has been completed, splash a final layer of plaster over the cast to give it extra strength and thickness, and rule it off.

Figure 6.58 Ruling off the final layer of plaster

STEP 11 Let the cast set for approximately 30 minutes then remove the timber fence and carefully lift the cast from the bench. Clean off the excess plaster with a joint rule.

Figure 6.59 Completed plain-faced slab

PRACTICAL TASK

4. MELTING AND POURING PROCEDURE WHEN USING HOT MELT COMPOUND

OBJECTIVE

To practise melting PVC in a hot melt machine, and pouring it onto a mould.

TOOLS AND EQUIPMENT

Hot melt machine   Timber for fence

Knife

PPE

Ensure you select PPE appropriate to the job and site conditions where you are working. Refer to the PPE section of Chapter 1.

STEP 1 Prepare the correct amount of PVC by cutting it into small pieces to aid the melting process.

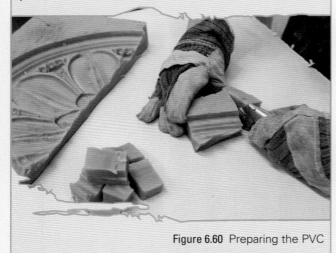

Figure 6.60 Preparing the PVC

STEP 2 Check the manufacturer data sheet regarding the grade of PVC you are using. This will give you the temperature needed to melt the material.

STEP 3 Place the melting pot under an extractor, to remove any fumes that are generated.

Figure 6.61 Position of the melting pot

STEP 4 Set the hot melt machine to the appropriate temperature, and allow the material to melt.

PRACTICAL TIP

The slower the material melts, the better the reproduction will be.

STEP 5 Prepare the model by pre-soaking with water, but ensure that there is no visible water on the surface.

PRACTICAL TIP

Excess water left on the model surface will turn into steam, leaving small pockets of bubbles on the surface.

Figure 6.62 The prepared model

STEP 6 Fix fences around the model and, when ready, pour the PVC in one continuous stream.

PRACTICAL TIP

Don't pour the PVC too slowly or directly onto the model, or seams will form.

Figure 6.63 Pouring the PVC

Figure 6.64 The finished mould

TEST YOURSELF

1. What metal would you normally use to make the template in a running mould?

   a. Nickel

   b. Zinc

   c. Bronze

   d. Aluminium

2. When would you attach a gig stick to a running mould?

   a. When you need to produce a circular moulding

   b. When you need to give the running mould extra stability

   c. When you need to produce a very long moulding

   d. When you need to make the running mould portable

3. What is the advantage of using super fine casting plaster?

   a. It is stronger than standard casting plaster

   b. It is cheaper than standard casting plaster

   c. It can reproduce very detailed patterns

   d. It is fire-resistant.

4. What is glue size?

   a. A release agent

   b. A retarder

   c. Plaster reinforcement

   d. A sealing compound

5. What sort of mould is formed by pouring moulding material over an open mould?

   a. A running mould

   b. A case mould

   c. A skin mould

   d. A flood mould

6. Which of these is the main advantage of using cold pour compound?

   a. It can be reused

   b. There are no health and safety risks

   c. It provides an accurate reproduction

   d. It is very cheap

7. What are the plaster mixes required for two-gauge casting?

   a. Fine and super fine

   b. Firstings and seconds

   c. Hot melt and cold pour

   d. Plaster of Paris and PVA

8. How should you treat laths before using them?

   a. Soak them in water

   b. Coat them in shellac

   c. Apply a thin coat of plaster

   d. Dust them with French chalk

9. What is hot melt compound made from?

   a. Tallow and paraffin

   b. Trisodium citrate

   c. Clay

   d. PVC

10. What are choppies used for?

    a. Smoothing the moulding after it is cast

    b. Reinforcing the plaster

    c. Cutting the running mould template

    d. Constructing a running mould

Chapter 7

Unit CSA–L2Occ55
FORM SAND AND CEMENT SCREEDS

LEARNING OUTCOMES

LO1/2: Know how to and be able to prepare for forming sand and cement screeds

LO3/4: Know how to and be able to prepare materials for forming sand and cement screeds

LO5/6: Know how to lay sand and cement screeds

INTRODUCTION

The aims of this chapter are to:

* describe the tools, equipment and materials you need to lay floor screeds

* explain the different types of floor screeds and when you would use them

* show you how to set out datum marks for levels and falls

* help you to mix screed material correctly

* show you ways of laying floor screeds to levels and falls

* explain how to compact, finish, cure and dry new floor screeds.

LAYING A FLOOR SCREED

Although there are specialist floor screeders, it is often the plasterer's job to lay a semi-dry **floor screed** after the walls and ceiling have been plastered. The purpose is to create a floor surface that is flat and smooth enough to be finished with tiles, carpet, laminate, floorboards or another floor covering. If the floor is not flat enough, the floor covering will not lie flat, which not only looks unattractive but could also cause people to trip or furniture to tip. It's therefore important to take time and care to get screeding right.

KNOW HOW TO PREPARE FOR FORMING SAND AND CEMENT SCREEDS

The hazards of forming sand and cement screeds

In Chapter 1 we examined the importance of identifying potential hazards and the ways in which risk assessments and method statements can help to avoid possible health and safety hazards.

The most common hazards you are likely to encounter when laying floor screeds are:

* screeding materials heating up when they are mixed, which could burn bare skin

* kneeling, crouching or bending for too long as you lay the floor, which could lead to back and knee problems

* poorly maintained and/or wrongly used power tools, which can result in injury

* on large sites, being hit by travelling plant or by something falling on your head

* poor housekeeping, for example tools being left around or cables crossing a room, resulting in slips, trips or falls

* injuries associated with manual handling.

It is also important to make sure that manufacturers' instructions are followed. These precautions ensure that you follow health and safety legislation and keep yourself safe.

PPE and protecting yourself from injury

Cement can burn your skin so always make sure that you cover up when working with it. Wear waterproof trousers instead of shorts, and long sleeves instead of T-shirts. Always wear gloves, goggles and overalls, and a face mask if you are dealing with fine, dusty particles. Wear a hard hat if necessary.

Preparing a floor involves a lot of bending and kneeling, which can put stress on your joints and result in long-term injuries to your knees and back. Kneepads or a knee mat distribute weight and pressure away from the knees, making it more comfortable to maintain a kneeling position. However, you still should get up regularly to walk or stretch.

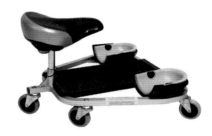

Figure 7.1 A kneeling creeper

Useful information sources about sand and cement screeds

As with any type of plastering, you must follow the specification provided for the job, along with any other information provided to you. Pay particular attention to drawings that show, for example, the required height of the floor. It is also important to note the type of floor covering (carpet, laminate, tiles etc.) that will be used on the screed, as this will affect your finish.

Several **British Standards** and European standards apply to the laying of floor screeds and the materials used.

PRACTICAL TIP

Always check the manufacturer's data sheet and the guidance printed on the bag of screed before you mix it. Floor screeds require a lot of material so, if you get it wrong, that's a lot of money wasted.

These include:

Code of standard	Name of standard	Contents of standard
BS 8000-9: 2003	Workmanship on building sites. Cementitious levelling and wearing screeds. Code of practice.	This standard provides recommendations and gives guidance on laying cementitious levelling screeds and wearing screeds, including on the quality of materials to be used.
BS 8204-1:2003+Amendment 1:2009	Screeds, bases and in situ floorings. Concrete bases and cementitious levelling screeds to receive floorings. Code of practice	This standard gives recommendations for the design and laying of concrete bases and cement levelling screeds to receive flooring.
BS EN 13318	Screed materials and floor screeds – Definitions	This standard provides agreed definitions for the technical terms being used in BS EN 13813, in English, French and German, so that they are clearly understood across the whole of Europe.
BS EN 13813	Screed materials and floor screeds – Screed material, Properties and requirements	This is the product standard that describes the essential characteristics of flooring products, specifies the methods by which these characteristics are to be determined and defines minimum levels of acceptable performance.
BS EN 13892	Methods of test for screed materials	A suite of eight test method standards for things like wear resistance and surface hardness.

Table 7.1 British and European standards relating to floor screeding

Tools and equipment for floor screeding

Some of the tools and equipment you will use will already be familiar from other plastering tasks. Others are used especially for floor screeding. Table 7.2 describes the main ones.

Tool / equipment	Description
Bucket	Buckets need to be big enough to contain the volume of your mix, and completely clean before you use them.
Builder's square	This is used to set out right angles – in floor screeding, this is when you are positioning the screed from the data point. It is wooden and braced with one side longer than the other. Squares are made from 75 mm × 30 mm timber and are half jointed at the 90° angle with a diagonal brace. Modern builder's squares are also often made from aluminium and can fold up.
Cement pump	This is used to transport large quantities of screed to where it is required. For example, if your screed is mixed away from where it is going to be laid, such as outside the building, a cement pump makes it easier to move it to where it is needed. The screed needs to be wet enough to flow easily through the hose. As with any electric equipment, you should not use a cement pump unless you have been trained to do so.
Chalk line	This is used to transfer floor thickness datum points around the area to be screeded. It is a reel of string into which you pour coloured powdered chalk, so that it transfers a straight line onto the surface as you reel it out. Some models can also be used as a plumb bob – the heavy end is dangled from the string to identify a vertical line.

Tool / equipment	Description
Float	This can be made out of wood or plastic and is used for smoothing and filling in hollows after you have ruled in the screed.
Floor laying trowel	This is used for laying, mixing and spreading flooring materials like sand and cement, and for giving the screed a smooth finish. It is made from steel and is either 400 mm or 455 mm long, has a pointed front and is normally much larger than the plastering trowel.
Gauging trowel	This is used to mix small amounts of material and to get plaster into difficult places, like corners of floor screeds. It can also be used to clean down other tools after screeding. It is made of steel and usually has a wooden or plastic handle.
Measuring tools	These include a long tape measure and a steel rule for measuring the thickness of the screed.
Mixer	A self-contained mixer is useful for mixing screed on larger jobs. Portable cement mixers have a capacity of 60 to 150 litres. They are powered by electricity (110V or 230V) or petrol.
Shovel	Shovels must be the right size for your height and strength. This is because you can easily strain your back when working with a loaded shovel.
Straight edge	This is a long, straight piece of wood or aluminium for levelling in the screeds and for ruling them off. They are usually about 3 m long, although they are available in shorter lengths, and about 100 mm wide and 35 mm deep. It's best to buy one rather than try to make one yourself, as it's important that the edges are completely parallel and the faces of the depth are flat.
Wheelbarrow	Wheelbarrows reduce the time and strain involved in transporting many heavy bags or loose sand or cement. This is essential when screeding which uses up a lot of material. Choose a wheelbarrow that is the right size for the job you are doing, e.g. ensure it will fit through doors and go around corners. Try to keep it clean so that the materials don't get contaminated.

Table 7.2 Tools and equipment used for floor screeding

You will also need to use levels to ensure that the floor is flat and to mark datum points. Table 7.3 describes the main types you will come across on site.

Type of level	Description
Laser level	This level sends out a laser beam to a target receiver. The level at that point can then be recorded for setting out. Models range from large industrial to small auto-setting versions. (See Fig 7.11.)
Automatic liquid level	This consists of a container of coloured liquid which is placed at a high point and attached with a hose to a metal gauge with an air valve. The datum level is set on the scale with a locking screw and water is drawn up to that level. The datum level can then be marked on the wall. (See Fig 7.12.)
Spirit level	This is a simple straight edge that has a glass tube containing a liquid and a bubble of air. When the air bubble is in the centre of the tube, the straight edge is exactly level. The tubes are marked with lines to confirm that the edge is level.
Water level	This is a length of hose that has a transparent tube in each end. The hose is filled with water. It is an ideal resource for checking levels in distances of over 30 m. The water level is a simple device, as water will always find its own level. One of its advantages is that you can take it around corners or obstructions. Note that these days it is not often used for levelling in new buildings.

Table 7.3 Types of level

Figure 7.2 Ordinary Portland cement

Figure 7.3 Sand suitable for floor screed

Materials and resources used to form sand and cement screeds

Materials for floor screeding usually contain 1 part ordinary Portland cement to 3 or 4 parts sand/aggregates. You can also get ready-mixed screed materials, often containing retarders to slow the set. This means more time can be spent laying the screed and less time spent shovelling sand and cement into the mixer.

Cement

Cement is added to bind the mix – that is, to stop it from falling apart. It is important to choose the specified cement type, because variations in cement quality can affect the strength and curing of the screed. The type of cement used for floor screeds is called **ordinary Portland cement**, which comes in different types according to the type of screed you need. Normally, you would use Portland Class 42.5N, to produce a fine concrete screed.

The initial set of a mix using ordinary Portland cement should not be less than 45 minutes and the final set not more than 10 hours.

Sand

The size and shape of aggregates affects how well the screed performs. The aggregate type specified for ordinary cement-sand levelling screed is washed sharp sand with particles of up to 4mm, as this makes it more hard wearing. Sand used for screeds should always be clean and well graded as it makes the mix more workable.

It is important to get the amount of sand in the mix right. Sand is cheaper than cement so it may be tempting to put in as much as possible – but the more sand in the mix, the weaker its strength, making it difficult to work. However, not enough sand (an under-sanded mix) will dry too quickly and the screed is likely to crack.

For general fine concrete screeds, you can replace 25 per cent of the sand with a single-sized aggregate, like granite chippings, using a mix of 1 part cement, 1 part aggregate and 1 part sand.

Do not use single-sized finely graded sand, fine brick-laying sands, crushed rock and sea sand containing shell.

Water

As when mixing any plaster, use enough clean, running water to work easily with your mix.

Polypropylene fibres

You can reinforce the screed and prevent it from cracking by adding polypropylene (PP) fibres. Beware that PVA (polyvinyl acetate) fibres are sensitive to water so should not be used in damp areas.

Admixtures

Admixtures can be added to ordinary Portland cement for different purposes, including:

* rapid-hardening

* retarding

* sulphate-resisting

* heat-resisting

* super-plasticising

* waterproofing/water repelling.

You may also add pigments to give it a particular colour.

Damp-proof membranes

To comply with Building Regulations, all new buildings must have a damp-proof membrane (DPM) installed under the floor screed to enable moisture-sensitive floor finishes, such as wooden floors or laminates, to be laid on top.

Most modern DPMs consist of a two-part epoxy mix, which is spread evenly across the floor as a liquid using a notched trowel or roller.

You can also use a thick polypropylene SPM sheet, though this will raise the level of the floor considerably so may not be appropriate in domestic settings.

Ready-mixed screeds

Many manufacturers offer pre-mixed screeds that contain additional materials, such as combining a DPM or to retard or speed up setting. Some 'green' varieties replace ordinary Portland cement with recycled materials. The manufacturers produce data sheets outlining their benefits and how to use them – always make sure you know exactly what the screed will do before you start to work with it.

Screed rails

If there is no curb or edging from which to work, a screed rail or trammel bar should be placed in the bedding material at the required level. This is a straight, sometimes L-shaped rod, usually 3 to 4.5m long. Many different types are available to buy, and may be made from uPVC, concrete, aluminium or steel. Some types are left in place to act as an expansion joint after the screed has set, while others are removed, or have a removable top.

Screed rails are firmly bedded onto the sub-base or laying course so that their top surface is at the level required for the surface of the screed. The

> **PRACTICAL TIP**
>
> Always make sure you follow the manufacturer's instructions, especially when using more than one admixture together.

Figure 7.4 A screed rail

level is determined by the compactibility of the laying course material and the thickness of the final floor covering. Rails need to be secure so that they do not move, settle or sag when the screed board is dragged over them, as this would mean your final screeded bed is at the wrong level. However, you still need to ensure that the rails can be removed, if necessary, before the final floor covering is applied.

The screed rails should be aligned at 90° to the direction of the screed ruling, and spaced so that the screed board will span the gap between them with an overlap of around 200 mm.

Once they are in place, you need to set and check their level using a taut string line stretched between two known datum points, or with an automatic level.

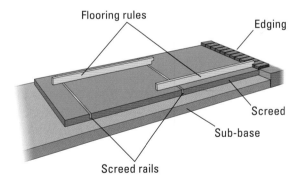

Figure 7.5 Positioning the screed rails

Various beads and trims

We saw the different uses for beads and trims in Chapter 5. Similar items are used in floor screeds to form edges, define corners and embed movement joints.

Cement-based materials shrink as they dry, as well as expanding and contracting with changes in the temperature, especially where there is underfloor heating. To reduce the screed cracking, movement joints are often specified, especially if the final floor covering will be rigid, such as ceramic tiles.

Calculating areas, volumes and ratios of floor screed materials

Take another look at Chapter 2 to refresh your memory on how to work out areas and volumes.

To calculate approximately how much material you will need, you need to know the:

* specified surface area of the screed

* specified depth of the screed

* mix ratio of the screed, e.g. 1:3 or 1:4 (see 'Materials and resources used to form sand and cement screeds', below)

* density of the materials used (as a rule of thumb, both sand and ordinary Portland cement are around 1.5 tonnes per cubic metre but this varies according to the brand).

If you want to use fast-drying screed, you may also need to know the drying time.

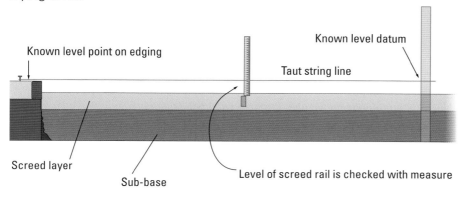

Figure 7.6 Checking the level of the screed rail

Labels: Known level point on edging; Known level datum; Taut string line; Screed layer; Sub-base; Level of screed rail is checked with measure

DID YOU KNOW?

BS 8204-2 provides guidance for installing and spacing joints in floor screed.

Densities

Sand	1.27 t/m³
Cement	1.7 t/m³

Dimensions

	Area 1	Area 2	Area 3	Area 4	Area 5	Area 6
Length in metres						
Width in metres						
Thickness in metres						
	0	0	0	0	0	0

Total in m³ [0]

Mix 1 to 4

	Cement	Sand
M³	0	0
Kg (inc contingency)	0	0

Reset form

Figure 7.7 An example of an online materials calculator (*source* www.source4me.co.uk)

PRACTICAL TIP

The better the preparation, the sounder the floor.

Types of floor screed

There are four main types of floor screed:

- bonded/monolithic
- unbonded
- floating
- separate.

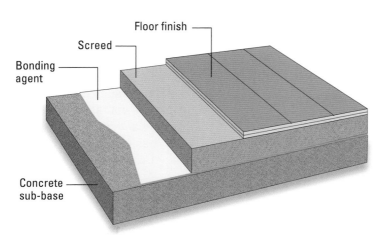

Figure 7.8 A bonded floor screed

Bonded

This is the most common type. A hardened concrete base is prepared using cement and PVC, and ruled but left rough to help the screed bond.

The floor screed is then laid on top of it. The floor should be laid to a minimum thickness of 25 mm and would preferably be 40 mm thick. If it is too thick, however, the floor will not bond properly.

Fully bonded screeds are less likely to crack or curl so are the best type to use where there is going to be heavy duty usage or rigid flooring.

Monolithic

This is a type of bonded floor screed that is placed within three hours of the concrete base being laid. This will create a particularly strong bond with minimal shrinking because the concrete and screed will dry slowly together. It is between 10 mm and 15 mm thick.

Figure 7.9 An unbonded bonded floor screed

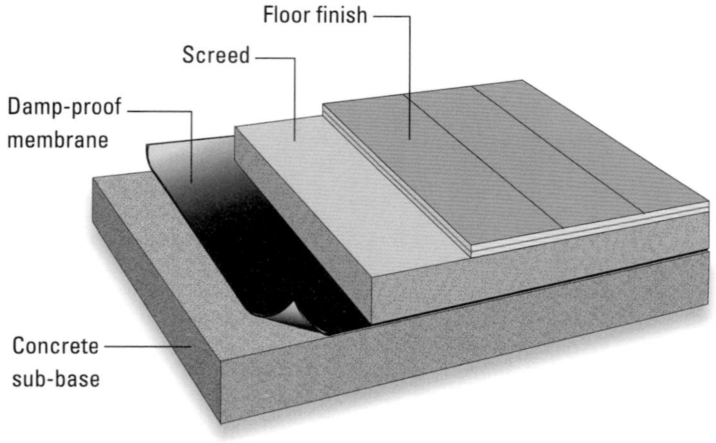

Unbonded

Here, the floor is not bonded directly to its concrete sub-base, perhaps because of the concrete's poor condition or because there is a damp-proof sheet membrane laid on the concrete before the screed is applied. Its minimum thickness is 50 mm. This sort of floor is suitable for underfloor heating (if laid to 65 mm) or anywhere that might be affected by damp rising through the floor.

Floating floor screed

This like an unbonded floor, but is laid over insulation boards to a thickness of 65 mm to 100 mm. It should be at the thicker end of the scale in commercial locations, such as shops and offices, but can be thinner in domestic locations like houses.

Separate

This is a normal concrete floor where the screed has been left to bond to the surface. It is 40 mm or more thick.

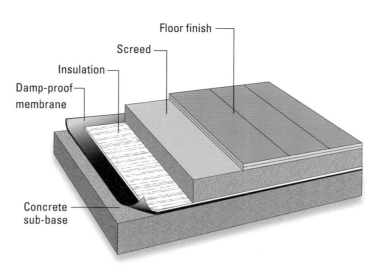

Floor finish
Screed
Insulation
Damp-proof membrane
Concrete sub-base

Figure 7.10 A floating floor screed

PREPARING MATERIALS FOR FORMING SAND AND CEMENT SCREEDS

Describe the method of laying floor screeds to levels and falls

Setting out levels

Whether the floor will be laid level or to a **fall**, the first task you must do is to strike a datum line around the area. Usually floors are laid level but sometimes they must be laid to a fall. In Chapter 2, we looked at using datum points as a reference for creating a level area. On a construction site, this should already have been worked out and marked. If this has not happened, or if you are at a private house, you will need to work out your own datum level, for example the bottom of a door. This is marked and then transferred around the building to wherever it is needed.

Whether the floor will be laid level or to a fall, you first need to mark the datum points at regular intervals around the room and then join them up using a chalk line.

Using a laser level

This is the quickest and easiest method. The laser beam is transmitted across the room as a red dot and you simply mark where it is before joining up your marks.

Figure 7.11 A laser level in use

REED TIP

Good communication is about listening – not just talking – and understanding the different ways people prefer to communicate.

213

Using an automatic liquid level

One way to form the datum line is by using an automatic liquid level. You must put the container of water at a height that is compatible with working areas for setting levels and falls; neither so low that it will get in the way of the floor, nor so high that you can't use it accurately.

Set the gauge on the rod to zero at the liquid level of the container, then move the rod to a point where you want to make a datum mark. Move the rod up or down until the liquid level is at the zero point you set on the gauge, then transfer this level by making a pencil mark on the wall. Repeat around the room then join up all the marks with a chalk line.

PRACTICAL TIP

If possible, set it on the floor, on a stool, on a box, on a pair of steps or on a high cupboard, such as a wall unit.

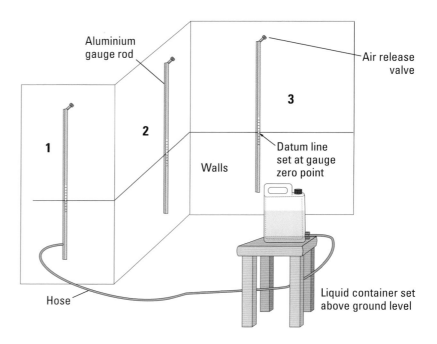

Figure 7.12 Using an automatic liquid level to mark datum points

Using a water level

This works on the same principle as the automatic liquid level but is simpler. Fix one end so that the water is at the height of the datum point. Put the other end where you want the second datum mark and wait until the water has settled. It finds its own level so will be at the same height as the first end. Mark this height on the wall and repeat around the room.

Using dots to level the screed

In Chapter 5 we saw how to screed a wall using dots. The same principle can be used for creating a level floor. Start by applying about half the required thickness of screed to a width of about 300 mm. Compact it using the float to make sure all the material is pressed together and firm enough to receive a dot. It is best to make this dot fairly long. Then lay a second layer of screed and use a flooring rule to rule it in. Place a small timber batten or screed rail on it to the height you need. If a thick floor covering is going to be used over the screed, such as tiles or floorboards, place one of these on top of the batten to get the final floor height. Use a builder's square to check the levels of the screed against the datum line or door frame.

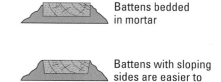

Battens bedded in mortar

Battens with sloping sides are easier to remove

Figure 7.13 Batten placement in dots

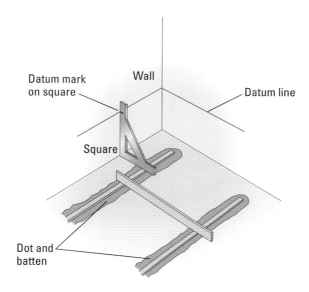

Figure 7.14 Levelling screed floors using dots and battens

Once you are satisfied that the first dot is accurate, set more dots at intervals around the room. When you set each one, check it is level by placing a long piece of timber between the dot and the previous dot and putting a spirit level on the timber. When you have placed dots all around the room, check again that they are level to within 3 mm over a 3 m distance.

Preparing floors with a fall

The dots method is especially useful for preparing floors with a fall, as you can set the dots at different heights. Set one dot onto the lowest part of the floor and then fix the rest of the dots to the line to give it a regular decline.

Another way is to use your flooring rule to help you. For example, if you need to lay a fall of 1:100, this is the same as 10 mm in every metre. So if the flooring rule or straight edge is 3 m long, place a dot or batten at 30 mm below one end of it. When you level the straight edge using a spirit level, this will mean that from one dot to another the floor will rise by 30 mm.

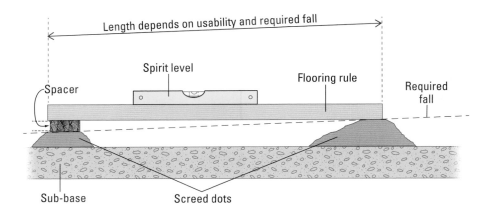

Figure 7.15 Using screed dots to achieve a fall

Gauging and mixing materials to the correct consistency

Most floor screeding contractors will tell you that getting the mix right is the most important element of the job. If it is not mixed correctly, the laying and finish of the screed will not be of good enough quality and the floor will be weak.

Gauging screed materials

The ratio of sand to cement is usually 3:1 – that is, 3 parts sand to 1 part cement. Sometimes 1:4 or 1:4.5 may be specified, and this is often the proportion used in ready-mixed screeds. Check this before you start to mix.

As we saw in Chapter 4, the easiest way to gauge the correct amounts is to fill a bucket with sand or cement, level it off and mix on the floor or in the mixer. To get a ratio of 3:1, you'd need 3 buckets of sand and 1 of cement.

Mixing screed materials

Floor screed material should be drier and more powdery than wall plasters and renders. However, you don't want it too dry as this will prevent a good finish. You also don't want it to be too wet, as this will affect the time the screed takes to dry and cause the cement to rise to the surface, which will weaken the floor. A wet mix will also create poor workability. The trick is therefore to add the correct amount of water.

Whether you are mixing by hand or in the mixer, add a little water at a time until the mix is damp. You can always add more water if you need to, but if it's too wet you'll need to add more materials, which is expensive and could affect the proportions.

If you are mixing by hand, make a well in the middle of your pile of materials and pour water into it. Then bring in the materials bit by bit with your shovel and mix them with the water.

If you using a mixer, put in a small amount of water first then add some sand and finally some cement. Repeat, adjusting the amounts as necessary until you have the correct consistency.

Figure 7.16 The snowball test

> **PRACTICAL TIP**
>
> Check your screed consistency with the 'snowball test'. Take a handful of the screed mix in your gloved hand and squeeze it. The screed should hold together without crumbling or dripping and should not stick to your hand.

> **PRACTICAL TIP**
>
> If a cement pump is going to be used, first check on the manufacturer's instructions that the mix can be used with a pump, and that you know what consistency is needed. The mix often needs to be wetter so that it can travel through the hose.

Table 7.4 describes some of the effects of not gauging and mixing correctly.

Problem	Possible effect
Not enough cement	The floor will be too weak so may crack or not be able to take a load.
Not enough sand	The mix will not bind properly, so will shrink, crack and fall apart.
Mix too dry	The screed is harder to lay. The finish will be poor. The floor will be weak.
Mix too wet	The screed is harder to lay. The screed takes longer to dry. The finish will be poor. The floor will shrink too much as it dries, and cement may rise to the surface, making it weak.
Insufficient mixing	The screed will be lumpy, preventing a flat, level surface and giving a poor finish. Clumps of sand prevent the mix from binding. Clumps of cement make the mix sticky and inconsistent.

Table 7.4 Effects of incorrect gauging and mixing

Protecting the work and its surrounding area from damage

Newly laid screeds should be protected from impact and friction as damage to the screed could affect the quality of the final flooring. Remember that screeds are not the finished floor so are not intended to be used as a permanent surface.

The British Standards state that screeds should be protected from direct traffic (such as people walking over them, tools being placed on them and plant driving over them) as soon as possible after installation. Measures could involve clear signage, redirecting site traffic on large construction sites and installing screed protectors.

Figure 7.17 Damaged floor screed

PRACTICAL TIP

Screed protectors are permeable plastic or polypropylene mats that enable moisture to escape from the drying screed. They are flexible and lightweight, so won't themselves damage the screed. Some are installed above the screed so can be walked on.

Protective measures should only be removed when the final floor covering is about to be installed. It is important to pay attention to the weight loaded on the screed and the level of compression of any insulation installed under the screed.

DID YOU KNOW?

Failing to protect a screed could not only damage it but also void any warranty on it. This should be made clear to clients.

The time taken before the screed can take foot traffic depends on the brand and type of screed used. Generally it is between 2 and 48 hours – the exact time should be stated on the packaging or manufacturer's data sheet. Care still needs to be taken as the screed might not completely dry for weeks.

DID YOU KNOW?

On some large sites, mobile elevating working platforms (MEWPs) travelling over the curing screed have squashed the insulation, which has ruined the screed. This can be avoided by checking all the specifications and other information at the start of the job.

LAYING SAND AND CEMENT SCREEDS

Laying screeds to levels and falls

Once you have prepared the sub-base, marked the datum levels and placed your dots, you are ready to lay the screed between the dots.

First brush grout onto the area you want to screed. The screed must be applied while the grout is still wet.

Whether the screed is poured, shovelled or pumped, you need enough to spread over a large enough area to be able to rule it off and compact it. This means spreading the screed to a thickness of about 10mm higher than the levelled edges.

Spread the screed with a flooring trowel or a shovel (for larger areas) and lightly compact it with a trowel or float.

Rule it off by pushing and pulling a flooring rule over the surface.

Figure 7.18 Ruling off the screed

Try to keep the pressure even, so that the width of the floor is levelled equally.

Infill any holes with screed and repeat until you are satisfied that the section is ruled level and flat. Remove the dots or screed rails if they show above the screed.

You can then finish the floor with a float or trowel (see page 222).

PRACTICAL TIP

Remember to start at the far end and work towards the door so that you don't get trapped in a corner and have to damage the screed by walking over it.

DID YOU KNOW?

If the screed is specified as being thicker than about 50mm, you will get best results by laying it in two stages. Lay the first half and compact it, then rake it to provide a key for the second half.

PRACTICAL TIP

Keep your hands a comfortable distance apart on the rule – 300mm to 500mm is best for most people.

PRACTICAL TIP

Finish as you go along, as you won't be able to reach over a large area without kneeling on levelled screed.

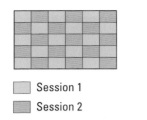

☐ Session 1
☐ Session 2

Figure 7.19 Bays for screeding

Using bays

A helpful system to avoid getting stuck or having to walk on the newly laid screed is to use alternate bays in what is known as a chequerboard pattern. This requires separating the floor into squares or rectangles (bays) with a width to length ratio of 1:1.5, and laying alternate bays so you can walk between them. Once the first set of bays is dry enough to walk and kneel on, you can come back to lay the rest of the bays.

The bays are made by dividing the room into sections using long timber battens embedded in screed. You could use your dots for this if you have laid them in an equal pattern. Fill in the spaces using a trowel and rule it off by resting your flooring rule across the battens.

PRACTICAL TIP

It can be tricky to get your rule into corners and edges of bays but it's important to ensure that these are as level as the centre of the floor.

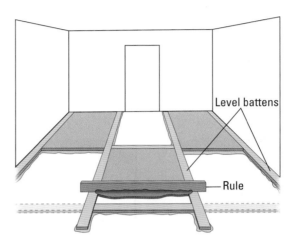

Figure 7.20 Screeding alternate bays

When you have filled in all the bays, remove the dots and battens then level them in, starting with the outside areas first. Remember to apply the finish as you work.

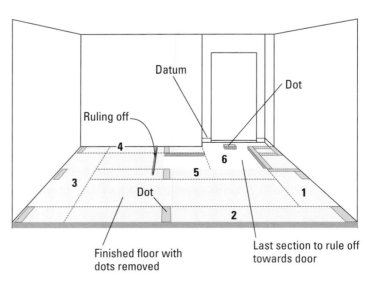

Figure 7.21 Rule and finish in the order shown

Laying floor to falls

Once you have established the levels, you can lay the falls. This might be a single slope downwards or all sides sloping towards a central drain or gulley. This drain should be fixed before you start to screed and, if you are using bays, these will be wedge-shaped if the falls are towards the centre.

Using self-levelling screeds

Self-levelling screeds are used to quickly and easily level or finish a floor that has minor cracks, lumps and hollows.

They can also be used as a finishing agent to give a smooth finish over traditional screed. Self-levelling screed is not a replacement for traditional screed as the overall quality of the finish is not generally regarded as being as good. However, it has several advantages:

* It is less labour-intensive.

* It can be laid to a shallower depth (generally 3 mm but cement-based self-levellers can be up to 40 mm).

* One bag covers up to 10 times more area.

* Shrinkage and curling are less likely to occur.

* Abrasion and impact resistance are better than for sand and cement floor screeds.

* It is less porous than concrete so will help with the adhesion of the finish.

There are disadvantages too:

* Self-levelling screed is not suitable where there are significant differences in height, such as large holes and bumps.

* You must work fast, as the mix dries very quickly.

* Despite its name, it takes some experience to get a level and smooth finish.

* It is very expensive, so you would need to balance the time saved against the cost of the materials.

Latex-based self-levellers are most commonly used, although you can also use water-based, acrylic or fibre-bonded types, depending on your sub-base. Self-levelling screed generally comes as a ready-mix to which you just add water. As with other types, the success of the final screed depends on it being mixed correctly.

It should be mixed more thinly than other types of floor screed, so that it flows across the floor. Add water and mix using an electric whisk until it's the consistency of tinned cream of tomato soup – try not to introduce air bubbles as these will remain when it dries.

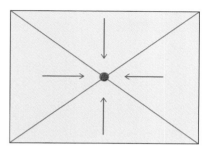

Figure 7.22 Work towards the centre of the fall

PRACTICAL TIP

You still need to grind down any high spots and fill in dips with mortar, to get a surface level to within 3 mm per metre.

PRACTICAL TIP

You may need to apply a DPM or prime the floor first with PVA.

Pour a small amount of the self-levelling compound onto the floor, starting with the lowest points, and use a trowel or spiked roller to join up the gaps. You don't need to get a perfect finish – as long as you have covered the area before the screed dries, it should set to create a reasonably smooth surface on which to lay the floor finish.

Compacting and finishing screeds

Compacting

Compacting the screed helps it to bond and stops it from falling apart. For smaller areas or in bays, you would do this with a float or trowel, but you can use a roller for larger areas. You need to apply the finish straight after you have compacted the area, before the screed starts to dry, to provide the appropriate key for the final floor covering.

Finishes

Floors are generally finished with either a floated or trowelled surface – whichever is suitable for the floor covering that is going to be laid. A wood float produces a sand-faced texture while trowelling gives a smooth finish which closes in the surface of a wood float finish.

Float finish

A large float is used to finish the surface if it is going to be covered with a heavy finish such as floorboards, mastic asphalt, asphalt tiles, bitumen or concrete tiles.

Trowel finish

A flooring trowel is suitable where a smooth surface is important, for example when the floor will be covered with carpet, hardboard, vinyl or cork tiles.

You can also use a latex self-levelling compound to give a smooth surface, as described above.

PRACTICAL TIP

You can also use power floats, tampers and rollers.

Figure 7.23 Float finish

Figure 7.24 Trowel finish

Curing screeds

If screed dries out too quickly, it is likely to curl and crack. The process can be slowed down by laying polythene sheets, building paper or damp hessian sacks over the screeded area to retain moisture during the setting process. The sheets should be left on the surface for a curing time of five to seven days before being removed to provide airflow.

Another option is to spray a chemical curing agent over the floor, which breaks down over a week or so as it assists the cure.

After this curing period, the floor should be allowed to dry naturally. Even if it looks dry, it could take anything from a few hours to 45 days for a newly laid screed to be dry enough to take the final floor covering (or even over 100 days for traditional screeds that have been laid thickly). Installing the floor covering too soon could spoil the whole floor, so it is best to conduct a moisture test using a moisture meter to ensure the screed has dried.

DID YOU KNOW?

One rule of thumb is that it takes 1 day for every 1 mm thickness of screed to dry, and 2 days per millimetre beyond a thickness of about 40 mm. However, the actual time will depend on the type of screed used and the temperature and humidity of the area where it has been laid. Fast-drying super-plasticisers dry at a rate of about 3 mm per day.

If your screed needs to be dry more quickly, for example if schedules are tight, use a self-levelling variety and consider running a dehumidifier to remove moisture from the screed. However, trying to speed up the process may not result in an adequate quality of finish.

Setting up drainage outlets in screeds

Wet rooms and showers may require you to incorporate a drainage outlet into the screed. You must lay any waterproofing and the screed to the correct fall to ensure that it is level with the outlet. If the floor is going to be tiled, make an allowance for the extra height by finishing the screed below the outlet or grate.

Remember that floor with embedded drainage outlets or gulleys usually need to be laid to a fall. Consult the specifications for details about the angle of slope that is required.

CASE STUDY

South Tyneside Council's
Housing Company

You need practical skills as well as theory

Billy Halliday is a team leader at South Tyneside Homes.

'I didn't get great qualifications at school and I didn't have to take a numeracy or literacy test when I started out 30 years ago. I was just told to get on with it! Now there's much more theory and you do need the basic numeracy and literacy skills – using a tape measure, filling in forms, reading toolbox talks and putting them into practice – but the most important thing is an enthusiasm for the trade and mastering the practical skills. It comes down to being practically minded, being able to get your head down and work quickly and accurately on site.

I'd say to people thinking about being an apprentice, don't feel you can't do it if you don't have an A in maths. If you've got the right attitude, you'll learn on the job.'

BE ABLE TO PREPARE FOR FORMING SAND AND CEMENT SCREEDS

PRACTICAL TASK

1. PREPARING THE SUB-BASE FOR A BONDED FLOOR SCREED

TOOLS AND EQUIPMENT

Brushes	Flooring trowel
Buckets	Scraper

PPE

Ensure you select PPE appropriate to the job and site conditions where you are working. Refer to the PPE section of Chapter 1.

STEP 1 Ensure the surface of the concrete sub-base is clean and free from dust, rubble, loose plaster and anything else that would cause the screed not to bond properly. Scrape and sweep away loose dirt and dispose of it according to your site rules.

Figure 7.25 Scraping the floor

STEP 2 Brush the concrete sub-base with enough clean water to cover it. Leave the water to soak in, preferably overnight.

STEP 3 Brush off any excess water.

STEP 4 Mix the grout at a ratio of 1 part sand to 1 part cement with enough water to make a thick but workable consistency.

STEP 5 Apply the grout to the damp concrete sub-base using a trowel, then brush it over to ensure a consistent covering.

PRACTICAL TIP

Smooth surfaces may need to be roughened to form a mechanical key.

Figure 7.26 Applying grout to the concrete sub-base

STEP 6 When the grout has set, add a thin layer of PVA over it.

STEP 7 The sub-base is now ready for the floor screed.

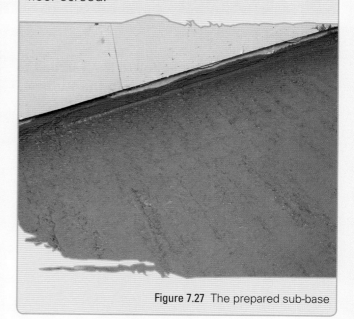

Figure 7.27 The prepared sub-base

2. PREPARING THE SUB-BASE FOR AN UNBONDED FLOOR SCREED OR FLOATING FLOOR

TOOLS AND EQUIPMENT

Brushes

Buckets

Flooring trowel

PPE

Ensure you select PPE appropriate to the job and site conditions where you are working. Refer to the PPE section of Chapter 1.

STEP 1 Ensure the surface of the concrete sub-base is clean and free from dust, rubble, loose plaster and anything else that would cause the screed not to bond properly. Sweep away loose dirt and dispose of it according to your site rules.

STEP 2 Ensure the sub-base is flat, smooth and free from cracks or hollows. If necessary, repair these using small amounts of grout or floor screed.

STEP 3 The sub-base is now ready for the damp-proof membrane or insulation boards and floor screed.

PRACTICAL TASK

3. FORM A FLOOR SCREED TO LEVEL

OBJECTIVE

To practise the skills needed to lay a screed to levels, and apply a float finish to the surface

INTRODUCTION

In this practical task, you will understand the types of floor screeds and materials used, and the quality of the finished surface. You will also understand the importance of datum levels and the need for accuracy in setting out.

TOOLS AND EQUIPMENT

Builder's square

Float

Flooring trowel

Shovel

Small piece of timber for dots

Spirit level

Straight edge

Water or laser level

PPE

Ensure you select PPE appropriate to the job and site where you are working. Refer to the PPE section of Chapter 1.

STEP 1 First check the floor is ready to receive the screed. Remove any loose particles and high points, and brush the whole area to remove the dust.

STEP 2 The next task is to strike a datum line around the area with a water level or a laser level. When you have levelled the lines on the wall around the room, join them together to form a datum line.

Figure 7.28 Checking the datum line

PRACTICAL TIP

If you are using the liquid level, make sure the container is at a height that is compatible with working areas for setting levels and falls (see page 214).

STEP 3 Now place the dots at 40 mm height from the floor, working from the datum line.

STEP 4 Start by applying about half the required thickness of screed to a width of about 300 mm. It is best to make this dot fairly long.

Compact it using the float so that when you place a small piece of timber on the screed, the timber and screed together are the required height of the floor. Then lay a second layer and use a flooring rule to rule it in. Place a small timber batten or screed rail on it to the height you need. Check the position of the dot with a straight edge and a level.

PRACTICAL TIP

Always work towards the door when laying a floor, so start from the furthest point away and ensure the dots are placed level.

STEP 5 Mix the floor screed according to the manufacturer's instructions or the specification if you are making it yourself. Using a shovel, add the material between the dots and compact it using a float until the screed is level with the dots. Now place the rule on the dots and, using them as a guide, rule towards you to fill in any hollows and rule it until the screed is straight and level with no uneven areas or hollows.

Figure 7.29 Ruling in the screeds

STEP 6 Once the screeds are completed, carry out the same process by filling in the main part of the floor area, using the screeds as a guide to rule off. Start to rub up the floor area with a float as you rule in, leaving a sand-faced texture, then trowel in the area to leave a smooth finish.

PRACTICAL TIP

Lightly trowel the area, taking care to not over-trowel as this will bring cement to the surface.

STEP 7 As you work towards the door, remove the dots and fill in any hollows.

Figure 7.30 The laid floor

PRACTICAL TASK

4. FORM A FLOOR SCREED TO A FALL

OBJECTIVE

To practise the skills necessary to lay a screed to a fall, and to apply a smooth trowel finish

INTRODUCTION

This method is used in wet areas like bathrooms and shower rooms, as you can set the dots at different heights.

TOOLS AND EQUIPMENT

Builder's square	Spirit level
Float	Straight edge
Flooring trowel	Timber for dots
Shovel	Water or laser level

PPE

Ensure you select PPE appropriate to the job and site conditions where you are working. Refer to the PPE section of Chapter 1.

STEP 1 Follow steps 1 and 2 from Practical Task 3.

STEP 2 Set one dot onto the lowest part of the floor using the method described in step 3 of Practical Task 3.

STEP 3 Fix the rest of the dots along in a line to give the floor a regular decline, depending on the fall required.

PRACTICAL TIP

For example, if you need to lay a fall of 1:100, this is the same as 10 mm in every metre. So if the flooring rule or straight edge is 3 m long, place a dot or batten at 30 mm below one end of it. When you level the straight edge

STEP 4 Place the drain or grid in position on the floor and ensure it is compacted into position.

STEP 5 Follow steps 4 and 5 in Practical Task 3, filling in the screeds to the correct height and ruling the floor towards the door area.

PRACTICAL TIP

Make sure you compact the material with a shovel and float, and apply a trowel finish.

STEP 6 Complete the floor with a float and smooth trowel finish.

TEST YOURSELF

1. What type of floor is laid on insulation boards?

   a. Monolithic floor

   b. Bonded floor

   c. Floating floor

   d. Separate floor

2. What sort of finish would you use if the floor was going to be covered with concrete tiles?

   a. Trowel finish

   b. Self-levelling

   c. No finish is required

   d. Float finish

3. What does DPM stand for?

   a. Damp-proof membrane

   b. Dots per metre

   c. Dust proof material

   d. Datum place mark

4. What sort of sand is best for use in floor screeds?

   a. Finely graded sand

   b. Sharp, well-graded sand

   c. Fine brick-laying sand

   d. Sea sand with shell particles

5. What is the classification of a level floor?

   a. + or – 30 mm in height over a 3 m distance

   b. + or – 1 mm in height over a 1.8 m distance

   c. + or – 3 mm in height over a 1 m distance

   d. + or – 3 mm in height over a 3 m distance

6. A screed rail must be…

   a. curved

   b. aluminium

   c. less than 200 mm long

   d. straight

7. What is the advantage of screeding in bays?

   a. You can walk between the bays so you don't have to stand on screed you have just laid

   b. It's quicker than laying screed over a large area

   c. Cement pumps can deliver the screed mix more accurately

   d. You can use a longer flooring rule to level the screed

8. When might a floor need to be laid to a fall?

   a. When the final floor covering will be wooden boards

   b. When stairs will be fixed to the floor

   c. When you are using a self-levelling screed

   d. When it requires a drainage gulley

9. Which of these is a disadvantage of self-levelling screed over traditional sand and cement screed?

   a. One bag doesn't cover as much area

   b. It is more likely to shrink

   c. It does not level out major defects in the sub-base

   d. It takes a long time to dry

10. How long should you leave a screed to cure under cover?

    a. About 100 days

    b. About a week

    c. 24 hours

    d. 45 days

INDEX